Encyclopedia of Optical Fiber Technology: Recent Advances

Volume I

Encyclopedia of Optical Fiber Technology: Recent Advances Volume I

Edited by **Marko Silver**

New York

Published by NY Research Press,
23 West, 55th Street, Suite 816,
New York, NY 10019, USA
www.nyresearchpress.com

Encyclopedia of Optical Fiber Technology: Recent Advances
Volume I
Edited by Marko Silver

International Standard Book Number: 978-1-63238-145-3 (Hardback)

Contents

Preface

This book provides an extensive analysis of the recent developments and progress made in optical fiber technology highlighting the newest range of optical communication, system and network, sensor, laser, measurement, characterization and devices. It gives emphasis to topics such as optical fiber communication systems and networks, and plastic optical fibers technologies. The chapters within this book have been contributed by prominent academicians and scientists involved in state of the art research in the field of photonics. This work will serve as a reference for readers from both academics and industrial backgrounds.

The researches compiled throughout the book are authentic and of high quality, combining several disciplines and from very diverse regions from around the world. Drawing on the contributions of many researchers from diverse countries, the book's objective is to provide the readers with the latest achievements in the area of research. This book will surely be a source of knowledge to all interested and researching the field.

In the end, I would like to express my deep sense of gratitude to all the authors for meeting the set deadlines in completing and submitting their research chapters. I would also like to thank the publisher for the support offered to us throughout the course of the book. Finally, I extend my sincere thanks to my family for being a constant source of inspiration and encouragement.

Editor

Optical Fiber Systems and Networks

Scaling the Benefits of Digital Nonlinear Compensation in High Bit-Rate Optical Meshed Networks

Danish Rafique and Andrew D. Ellis

Additional information is available at the end of the chapter

1. Introduction

The communication traffic volume handled by trunk optical transport networks has been increasing year by year [1]. Meeting the increasing demand not only requires a quantitative increase in total traffic volume, but also ideally requires an increase in the speed of individual clients to maintain the balance between cost and reliability. This is particularly appropriate for shorter links across the network, where the relatively high optical signal-to-noise ratio (OSNR) would allow the use of a higher capacity, but is less appropriate for the longest links, where products are already close to the theoretical limits [2]. In such circumstances, it is necessary to maximize resource utilization and in a static network one approach to achieve this is the deployment of spectrally efficient higher-order modulation formats enabled by digital coherent detection. As attested by the rapid growth in reported constellation size [3,4], the optical hardware for a wide variety of coherently detected modulation formats is identical [5]. This has led to the suggestion that a common transponder may be deployed and the format adjusted on a link by link basis to either maximize the link capacity given the achieved OSNR, or if lower, match the required client interface rate [6] such that the number of wavelength channels allocated to a given route is minimized. It is believed that such dynamic, potentially self-adjusting, networks will enable graceful capacity growth, ready resource re-allocation and cost reductions associated with improved transponder volumes and sparing strategies. However additional trade-offs and challenges associated with such networks are presented to system designers and network planners. One such challenge is associated with the nonlinear transmission impairments which strongly link the achievable channel reach for a given set of modulation formats, symbol-rates [6,7] across a number of channels.

Various methods of compensating fiber transmission impairments have been proposed, both in optical and electronic domain. Traditionally, dispersion management was used to suppress the impact of fiber nonlinearities [8,9]. Although dispersion management is appreciably beneficial, the benefit is specific to a limited range of transmission formats and rates and it enforces severe limitations on link design. Similarly, compensation of fiber impairments based on spectral inversion (SI) [10], has been considered attractive because of the removal of in-line dispersion compensation modules (DCM), transparency to modulation formats and compensation of nonlinearity. However, although SI has large bandwidth capabilities, it often necessitates precise positioning and customized link design (e.g., distributed Raman amplification, etc.). Alternatively, with the availability of high speed digital signal processing (DSP), electronic mitigation of transmission impairments has emerged as a promising solution. As linear compensation methods have matured in past few years [11], the research has intensified on compensation of nonlinear impairments. In particular, electronic signal processing using digital back-propagation (DBP) with time inversion has been applied to the compensation of channel nonlinearities [12,13]. Back-propagation may be located at the transmitter [14] or receiver [15], places no constraints on the transmission line and is thus compatible with the demands of an optical network comprising multiple routes over a common fiber platform. In principle this approach allows for significant improvements in signal-to-noise ratios until the system performance becomes limited only by non-deterministic effects [16] or the power handling capabilities of individual components. Although the future potential of nonlinear impairment compensation using DBP in a dynamic optical network is unclear due to its significant computational burden, simplification of nonlinear DBP using single-channel processing at the receiver suggest that the additional processing required for intra-channel nonlinearity compensation may be significantly lower than is widely anticipated [17,18]. Studies of the benefits of DBP have largely been verified for systems employing homogenous network traffic, where all the channels have the same launch power [19]. However, as network upgrades are carried out, it is likely that channels employing different multi-level formats will become operational. In such circumstances, it has been demonstrated that the overall network capacity may be increased if the network traffic will become inhomogeneous, not only in terms of modulation format, but also in terms of signal launch power [6,7,20]. In particular, if each channel operates at the minimum power required for error free propagation (after error correction) rather than a global average power or the optimum power for the individual channel, the overall level of cross phase modulation in the network is reduced [20].

In this chapter we demonstrate the application of electronic compensation schemes in a dynamic optical network, focusing on adjustable signal constellations with non identical launch powers, and discuss the impact of periodic addition of 28-*Gbaud* polarization multiplexed m-ary quadrature amplitude modulation (PM-mQAM) channels on existing traffic. We also discuss the impact of cascaded reconfigurable optical add-drop multiplexerson networks operating close to the maximum permissible capacity in the presence of electronic compensation techniques for a range of higher-order modulation formats and filter shapes.

2. Simulation conditions

Figure 1 illustrates the simulation setup. The optical link comprised nine (unless mentioned otherwise) 28-*Gbaud* WDM channels, employing PM-mQAM with a channel spacing of 50 GHz. For all the carriers, both the polarization states were modulated independently using de-correlated 2^{15} and 2^{16} pseudo-random bit sequences (PRBS), for x- and y-polarization states, respectively. Each PRBS was de-multiplexed separately into two multi-level output symbol streams which were used to modulate an in-phase and a quadrature-phase carrier. The optical transmitters consisted of continuous wave laser sources, followed by two nested Mach-Zehnder Modulator structures for x- and y-polarization states, and the two polarization states were combined using an ideal polarization beam combiner. The simulation conditions ensured 16 samples per symbol with 2^{13} total simulated symbols per polarization. The signals were propagated over standard single mode fiber (SSMF) transmission link with 80 km spans, no inline dispersion compensation and single-stage erbium doped fiber amplifiers (EDFAs). The fiber had attenuation of 0.2 dB/km, dispersion of 20 ps/nm/km, and a nonlinearity coefficient (γ) of 1.5/W/km(unless mentioned otherwise). Each amplifier stage was modeled with a 4.5 dB noise figure and the total amplification gain was set to be equal to the total loss in each span.

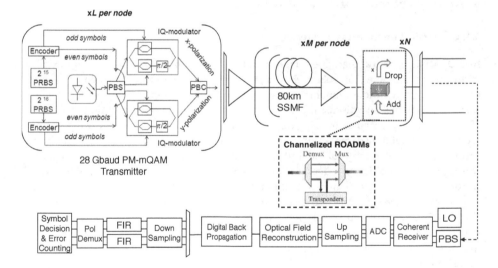

Figure 1. Simulation setup for 28-*Gbaud* PM-mQAM (m= 4, 16, 64, 256) transmission system with *L* wavelengths and *M* spans per node (total spans is given by *N*).

At the coherent receiver the signals were pre-amplified (to a fixed power of 0 dBm per channel), filtered with a 50 GHz 3rd order Gaussian de-multiplexing filter, coherently-detected and sampled at 2 samples per symbol. Transmission impairments were digitally compensated in two scenarios. Firstly by using electronic dispersion compensation (EDC) alone, employing finite impulse response (FIR) filters (T/2-spaced taps) adapted using a least mean square algorithm. In the second case, electronic compensation was applied via single-channel digital back-propagation (SC-DBP), which was numerically implemented by split-step Fourier method based solution of nonlinear Schrödinger equation. In order to establish the maximum potential benefit of DBP, the signals were up sampled to 16 samples per bit and an upper bound on the step-size was set to be 1 km with the step length chosen adaptively based on the condition that in each step the nonlinear effects must change the phase of the optical field by no more than 0.05 degrees. To determine the practically achievable benefit, in line with recent simplification of DBP algorithms, e.g. [17,18,21], we also employed a simplified DBP algorithm similar to [21], with number of steps varying from 0.5 step/span to 2 steps/span. Following one of these stages (EDC or SC-DBP) polarization de-multiplexing, frequency response compensation and residual dispersion compensation was then performed using FIR filters, followed by carrier phase recovery [22]. Finally, the symbol decisions were made, and the performance assessed by direct error counting (converted into an effective Q-factor (Q_{eff})). All the numerical simulations were carried out using VPItransmissionMaker®v8.5, and the digital signal processing was performed in MATLAB®v7.10.

3. Analysis of trade-offs in hybrid networks

3.1. Constraints on transmission reach

In a dynamic network, there are a large range of options to provide the desired flexibility including symbol rate [23], sub-carrier multiplexing [24], network configuration [25] signal constellation and various combinations of these techniques. In this section we focus on the signal constellation and discuss the impact of periodic addition of PM-mQAM (m= 4, 16, 64, 256) transmission schemes on existing PM-4QAM traffic in a 28-*Gbaud* WDM optical network with a total transparent optical path of 9,600 *km*. We demonstrate that the periodic addition of traffic at reconfigurable optical add-drop multiplexer (ROADM) sites degrades through traffic, and that this degradation increases with the constellation size of the added traffic. In particular, we demonstrate that undistorted PM-mQAM signals have the greatest impact on the through traffic, despite such signals having lower peak-to-average power ratio (PAPR) than dispersed signals, although the degradation strongly correlated to the total PAPR of the added traffic at the launch point itself. Using this observation, we propose the use of linear pre-distortion of the added channels to reduce the impact of the cross-channel impairments [26,27].

Note that the total optical path was fixed to be 9,600 *km* and after every *M* spans, a ROADM stage was employed and the channels to the left and right of the central channel were dropped and new channels with independent data patterns were added, as shown in Figure 2. in order to analyze the system performance, the dropped channels were coherently-detected after first ROADM and the central channel after the last ROADM link.

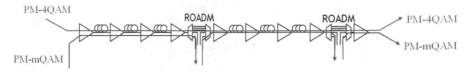

Figure 2. Network topology for flexible optical network, employing PM-4QAM traffic as a through channel, and PM-mQAM traffic as neighboring channels, getting added/dropped at each ROADM site. Note that in this schematic only right-hand wavelength is shown to be added/dropped, however in the simulations both right and left wavelengths were add/dropped. The total path length was fixed to 9,600 km, and the number of ROADMs was varied.

The optimum performance of the central PM-4QAM channel at 9,600 km occurred for a launch power of -1 dBm. In this study, the launch power of all the added channels was also fixed at -1 dBm, such that all channels had equal launch powers. Figure 3 illustrates the performance of the central test channel after the last node (solid), along with the performance of co-propagating channel employing various modulation formats after the first ROADM node (open) for a number of ROADM spacing's, using both single-channel DBP (Figure 3a) and EDC (Figure 3b). It can be seen that single-channel DBP offers a Q_{eff} improvement of ~1.5 dB compared to EDC based system. This performance improvement is strongly constrained by inter-channel nonlinearities, such that intra-channel effects are not dominant. Moreover, the figure shows that as the number of ROADM nodes are increased, or the distance between ROADMs decreases, the performance of higher-order neighboring channels improves significantly due to the improved OSNR.

It can also be seen from Figure 3 that added channels with higher-order formats induce greater degradation of the through channel. In particular if there are 30 ROADM sites (320 km ROADM spacing) allocated to transmit PM-64QAM, whilst this traffic operates with significant margin, the through traffic falls below the BER of 3.8x10^{-3}. This increased penalty is due to the increased nonlinear degradation encountered in the first span after the ROADM node, where higher formats induce greater cross phase modulation(XPM) than PM-4QAM by virtue of their increased PAPR. However, even when the add drop traffic is PM-4QAM, the performance of the through channel degrades slightly as the number of ROADM nodes is increased, despite the reduction in PAPR due to the randomization of the nonlinear crosstalk.

The estimated PAPR evolutions for the various formats are shown in Figure 4. Asymptotic values are reached after the first span, and reach a slightly higher value for m ≥ 16. The PAPR is reduced at the ROADM site itself, particularly for PM-4QAM. Figure 4 implies that harmful increases in the instantaneous amplitude of the interfering channels are not the entire cause of the penalty experienced by the through channel; we can therefore only conclude that the additional distortion results from interplay between channel walk off and nonlinear effects. Given that walk-off is known to induce short and medium range correlation in crosstalk between subsequent bits, effectively low pass filtering the crosstalk [28]. We thus believe that the penalty experienced by the through channel is not only because of variation in PAPR, but also due to the randomization of the crosstalk by the periodic replacement of the interfering data pattern.

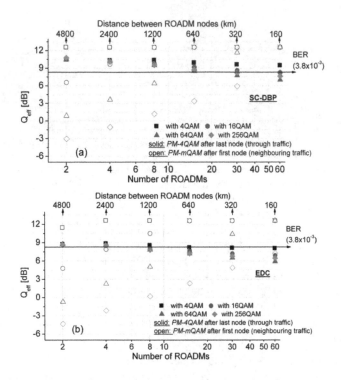

Figure 3. Q_{eff} as a function of number of ROADMs (and distance between ROADM nodes) for 28-*Gbaud* PM-mQAM showing performance of central PM-4QAM (solid, after total length), and neighboring PM-mQAM (open, after first node). a) with single-channel DBP, b) with electronic dispersion compensation. Square: 4QAM, circle: 16QAM, up triangle: 64QAM, diamond:256QAM. Up arrows indicate that no errors were detected, implying that the Q_{eff} was likely to be above 12.59 *dB*. Total link length is 9,600 km.

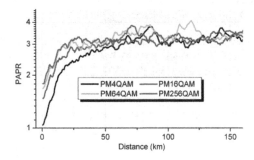

Figure 4. Variation in PAPR, for 4QAM (black), 16QAM (red), 64QAM (green) and 256QAM (blue) for a loss-less linear fiber with 20 ps/nm/*km* dispersion.

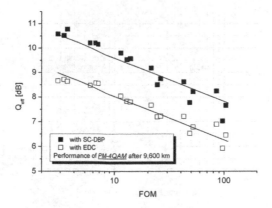

Figure 5. Q_{eff} of the PM-4QAM through channel for 28-*Gbaud* PM-mQAM add/drop traffic after 9,600 *km* as a function of a figure of merit (FOM) defined in the text for various add drop configurations. Solid: with single-channel DBP, open: with EDC.

This is confirmed by Figure 5, which plots the Q_{eff} of PM-4QAM after last node, for both EDC and single-channel DBP, in terms of a figure of merit (FOM) related to the increased amplitude modulation experienced by the test channel in the spans immediately following the ROADM node, defined as,

$$FOM_{PM-mQAM}(m) = (ROADM_N) \times \left[I_{max}(m) \middle/ I_{all}(m) \right] \qquad (1)$$

where m represents the modulation order, $ROADM_N$ represents number of add-drop nodes, I_{max} and I_{all} are the maximum and mean intensity of the given modulation format at the ROADM site. A strong correlation between the penalty and change in PAPR is observed. For instance, for a high number of ROADMs the system would be mostly influenced by relatively un-dispersed signals and the difference between peak-to-average fluctuations for multi-order QAM varies significantly. This leads to higher-order modulation formats impinging worse cross-channel effects on existing traffic for shorter routes.

Having observed that the nonlinear penalty is determined by the reduction in the correlation of nonlinear phase shift between bits arising from changing bit patterns, and to changes in PAPR arising from undistorted signals, it is possible to design a mitigation strategy to minimize these penalties. Figure 6 illustrates, for both EDC and single-channel DBP systems, that if the co-propagating higher-order QAM channels are linearly pre-dispersed, the performance of the PM-4QAM through traffic can be improved. The figure shows that when positive pre-dispersion is applied, such that the neighboring channel constellation is never, along its entire inter node transmission length, restored to a well-formed shape, the impact of cross-channel impairments on existing traffic is reduced significantly.

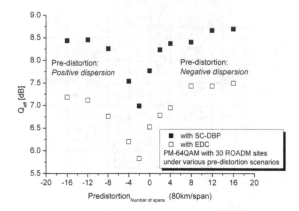

Figure 6. Q_{eff} of the PM-4QAM through channel with 30 ROADM sites, when the neighboring PM-64QAM channel is linearly pre-dispersed. Solid: with single-channel DBP, open: with EDC.

On the other hand, when negative pre-dispersion of less than the node-length (distance per node) is employed, the central test channel is initially degraded further. This behavior can be attributed to the increased impact of the PAPR of the un-dispersed constellation which is restored in the middle of the link. However, if negative pre-dispersion of more than the node-length is employed, the penalty is reduced due to lower PAPR induced XPM, and the performance saturates for higher values of pre-dispersion, similar to the case of positive pre-dispersion. Note that avoiding well formed signals along the entire link corresponds to max-imizing the path averaged PAPR of the signals. The benefits of this strategy have subsequently been predicted from a theoretical standpoint [27].

3.2. Constraints on transmitted power

In this section, we demonstrate that independent optimization of the transmitted launch power enhances the performance of higher modulation order add-drop channels but severe-ly degrades the performance of through traffic due to strong inter-channel nonlinearities. However, if an altruistic launch power policy is employed such that the higher-order add-drop traffic still meets the BER of 3.8×10^{-3}, a trade-off can be recognized between the per-formance of higher-order channels and existing network traffic enabling higher overall network capacity with minimal crosstalk [19].

As a baseline for this study, we initially consider transmission distances up to 9,600km with the same 80km spans, suitable to enable a suitable performance margin (at bit-error rate of 3.8×10^{-3}) for the network traffic given various modulation schemes at a fixed launch power of -1 dBm, (optimum power as determined in previous section. For a dynamic network with N ROADMs and m^{th} order PM-QAM, the overall results are summarized in Table 1. The ta-ble shows under which conditions the central PM-4QAM channel (right-hand symbol), and the periodically added traffic (left-hand symbol) are simultaneously able to achieve error-free operation after FEC. Two ticks indicate that both types of traffic is operational, whilst a cross indicates that at least one channel produces severely errorred signals. As expected, with decreasing ROADM spacing, the operability of higher-order neighboring channels increases due to the improved OSNR. However, it can also be seen that as a consequence, added channels with higher-order formats induce greater degradation of the through channel through nonlinear crosstalk as shown in Section 3.1. In particular, if the ROADM spacing is 320 *km*, allocated to transmit PM-64QAM, whilst this traffic is operable, the through traffic falls below the BER threshold. Conversely for large ROADM spacing, there is little change in nonlinear crosstalk, since the m-QAM signals are highly dispersed, but the higher order format traffic has insufficient OSNR for error free operation. We refer to this approach as "fixed network power".

mQAM/ ROADM spacing	4800 km	2400 km	1200 km	640 km	320 km	160 km	80 km
4QAM	++	++	++	++	++	++	++
16QAM	X+	++	++	++	++	+X	+X
64QAM	X+	X+	X+	++	+X	+X	+X
256QAM	X+	X+	X+	X+	XX	XX	+X

Table 1. Operability of PM-mQAM/4QAM above BER threshold of 3.8x10⁻³ for a total trnamsission distance of 9,600km. Tick/Cross (Left) represents performance of mQAM, Tick/Cross (Right) represents corresponding performance of central 4QAM. Tick: Operational, Cross: Non-operational

Since higher-order modulation formats have higher required OSNR, we expect the optimum launch power for those channels to be different than those used in the fixed network power scenario which was operated at a launch power of -1 *dBm*. Thus, for example, for large ROADM spacing, we improved performance might be expected if the add-drop traffic operates with increased launch power. Figure 7 illustrates the performance of through channel and the higher-order add-drop channels as a function of launch power of the add-drop traffic (through channel operates with a fixed, previously optimized, launch power of -1 *dBm*). For clarity we report two ROADM spacings, selected to give zero margin (Figure 7a) or ~2 *dB* margin (Figure 7b) for 256QAM add drop traffic. The ROADM spacing for 16 and 64QAM signals were scaled in proportion (approximately) to their required OSNR levels under linear transmission. The exact ROADM spacing is reported in the figure captions.

Figure 7 clearly illustrates that the higher-order formats operating over a longer (shorter) reach enable lower (higher) Q_{eff}, but also that the nonlinear effects increase in severity as the modulation order is increased. In particular, the long distance through traffic is strongly degraded before the nonlinear threshold is reached for such formats. Comparing Figure 7a and Figure 7b, we can see that the reduced ROADM spacing in Figure 7b enables improved performance of the add-drop channels; however the degradation of the through channel is increasingly severe. This change in behavior between formats can be attributed to the increased amplitude modulation imposed by un-dispersed signals added at each ROADM site, as discussed previously.

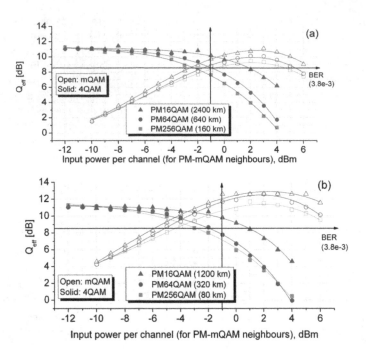

Figure 7. Q_{eff} as a function of launch power of two neighboring channels for 28-*Gbaud* PM-mQAM, showing perform-ance of central PM-4QAM (Solid), and neighboring PM-mQAM (Half Solid). Triangle: 16QAM, Circle: 64QAM, Square: 256QAM. The launch power per channel for PM-4QAM is fixed to -1 *dBm*. ROADM spacing of, a) 2400, 640, 160 *km*, b) 1200, 320, 80 *km* for 16, 64, 256 QAM, respectively.

We can use the results of Figure 7 to analyze the impact of various power allocation strat-egies. Clearly if we allow each transponder to adjust its launch power to optimize its own performance autonomously, a high launch power will be selected and the degradation to the traffic from other transponders increases in severity, and in all six scenarios in Figure 7 the through channel fails if the performance of the add drop traffic is optimized independently. This suggests that launch power should be centrally controlled. Howevercentrally control-led optimization of individual launch powers for each transponder is complex; so a more promising approach would be a fixed launch power irrespective of add-drop format or reach to minimize the complexity of this control. We have already seen (Table 1) that if the launch power is set to favor the performance of PM-4QAM (-1 *dBm*) the flexibility in trans-mitted format for the add/drop transponders is low, and to confirm this in Figure 7 four of the scenarios fail. The best performance for these two scenarios is achieved at a fixed launch power of -3 *dBm*, but we still find that 3 scenarios fail to establish error free connections. However, if the transponders are altruistically operated at the minimum launch power re-quired for the desired connection (not centrally controlled), the majority of the scenarios studied result in successful connections. The one exception is the add-drop of 256QAM channels with a ROADM spacing of 160 *km*, which is close to the maximum possible reach of the format. Note that shorter through paths would tend to use higher-order formats for all the routes, where nonlinear sensitivity is higher [29], and therefore we expect similar conclusions.

4. Application in meshed networks

In the previous section, we identified that optimum performance for a given predetermined modulation format was obtained by using the minimum launch power. However, this arbitrary selection of transmitted format fails to take into account the ability of a given link to operate with different formats, leading to a rich diversity of connections. In this section, we focus on the impact of flexibility in the signal constellation, allowing for evolution of the existing ROADM based static networks. We consider a configuration where network capacity is increased by allowing higher-order modulation traffic to be transmitted on according to predetermined rules based on homogenous network transmission performance. In particular we consider a 50 GHz channel grid with coherently-detected 28-*Gbaud* PM-mQAMand 20 wavelength channels. We demonstrate that even if modulation formats are chosen based on knowledge of the maximum transmission reach aftersingle-channel digital back-propagation, for the network studied, the majority of the network connections (75%) are operable with significant optical signal-to-noise ratio margin when operated with electronic dispersion compensation alone. However, 23% of the links require the use of single-channel DBP for error free operation. Furthermore, we demonstrate that in this network higher-order modulation formats are more prone to impairments due to channel nonlinearities and filter crosstalk; however they are less affected by the bandwidth constrictions associated with ROADM cascades due to shorter operating distances. Finally, we show that, for any given modulation order, a minimum filter Gaussian order of ~3 or bandwidth of ~35 *GHz* enables the performance with approximately less than 1 *dB* penalty with respect to ideal rectangular filters [30].

4.1. Network design

To establish a preliminary estimate of maximum potential transmission distance of each available format, we employed the transmission reaches identified in Section 3. These are suitable to enable a BER of 3.8×10^{-3} at a fixed launch power of -1 *dBm* assuming the availability of single-channel DBP. These conditions gave maximum reaches of 2,400 *km* for PM-16QAM, 640 *km* for PM-64QAM and 160 *km* for 256 QAM. Note that only single-channel DBP was considered in this study since in a realistic mesh network access to neighboring traffic might be impractical. WDM based DBP solution may be suitable for a point to point submarine link or for a network connection where wavelengths linking the same nodes co-propagate using adjacent wavelengths. Implementation of this condition would require DBP aware routing and wavelength assignment algorithms. This approach could enable significant Q_{eff} improvements or reach increases. For 64QAM, up to 7 dBQ_{eff} improvements were shown in [29], although the benefit depends on the number of processed channels [31].

We then applied this link capacity rule to an 8-node route from a Pan-European network topology (see highlighted link in Figure 8). To generate a representative traffic matrix, for each node, commencing with London, we allocated traffic demand from the node under consideration to all of the subsequent nodes, operating the link at the highest order constellation permissible for the associated transmission distance, and selecting the next wavelength. We note that none of the links in this chosen route were suitable for 256QAM, indeed only the Strasberg to Zurich and Vienna to Prague links are expected to be suitable for this format.

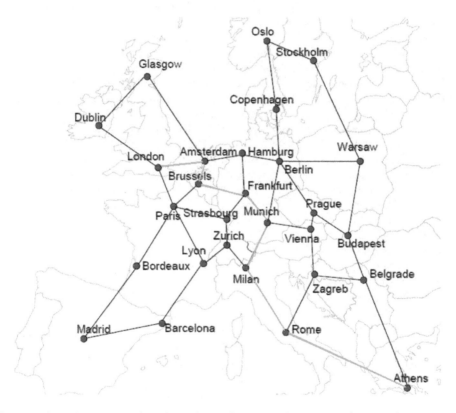

Figure 8. node Pan-European network topology. Link 1: London-to-Amsterdam: 7 spans, Link 2: Amsterdam-to-Brussels: 3 spans, Link 3: Brussels-to-Frankfurt: 6 spans. Link 4: Frankfurt-to-Munich: 6 spans, Link 5: Munich-to-Milan: 7 spans, Link 6: Milan-to-Rome: 9 spans, Link 7: Rome-to-Athens: 19 spans. (80 km/span).

Once all nodes were connected by a single link, this process was repeated (in the same order), adding additional capacity between nodes where an unblocked route was available until all 20 wavelengths were allocated, and no more traffic could be assigned without blockage.

Table 2 illustrates the resultant traffic matrix showing the location where traffic was added and dropped (gray highlighting) and the order of the modulation format (numbers) carried wavelength (horizontal index) on each link (vertical index). For example, emerging from node 6 are nine wavelengths carrying PM-4QAM and 5 wavelengths carrying PM-16QAM whilst on the center wavelength, PM-16QAM data is transmitted from node 1 (London) to node 5 (Munich) where this traffic is dropped and replaced with PM-64QAM traffic destined for node 6 (Milan). This ensured that various nodes were connected by multiple wavelengths. As it can be seen, the adopted procedure allowed for a reasonably meshed optical network (36 connections) with shortest route of 3 spans and longest path of 57 spans, emulating a quasi-real traffic scenario with highly heterogeneous traffic. At each node, add-drop functionality was enabled using a channelized ROADM architecture where all the wavelengths were de-multiplexed and channels were added/dropped, before re-multiplexing the data signals again. We considered Rectangular and Gaussian-shaped filters for ROADM stages, and the order of the Gaussian filters was varied from 1 through 6.

X (λ) / Y (Link)	-10	-9	-8	-7	-6	-5	-4	-3	-2	-1	0	+1	+2	+3	+4	+5	+6	+7	+8	+9
1	4	16	16					64	4	16	16	4	64		16	16	16		16	4
2	4	16	16	16	64	64	4	16	4	16	16	4	16	4	16	16	16		16	4
3	4	16	16	16	4	16	4	16	4	16	16	4	16	4	16	16	16	64	16	4
4	4	16	4	16	4	16	4	16	4	16	16	4	16	4	16	16	64	16	16	4
5	4	16	4	16	4	16	4	16	4	16	64	4	16	4	4	16		16		4
6	4		4		4		4	16	16	4	16	4	4	16		16				4
7	4		4		4		4		4	16	16		4							

Table 2. Traffic matrix (Each element represents the modulation order, Grayed: Traffic dropped and added at nodes highlighted in gray.

4.2. Results and discussions

4.2.1. Nonlinear transmission with ideal ROADMs

Figure 9 depicts the required OSNR of each connection as a function of transmission distance, after electronic dispersion compensation. Note that in this case we employed rectangular ROADM filters to isolate the impact of inter-channel nonlinear impairments from filtering crosstalk (no cascade penalties were observed with ideal filters).

Numerous conclusions can be ascertained from this figure. First, these results confirm that with mixed-format traffic and active ROADMs, as the transmission distance is increased the required OSNR increases irrespective of the modulation order due to channel nonlinearities.

Second, as observed by the greater rate of increase in required OSNR with distance, the higher-order channels are most degraded by channel nonlinearities, even at the shortest distance traversed. Furthermore, even for the shortest distances the offset between the theoretical OSNR for a linear system and the simulated values are greater for higher order formats. These two effectsare attributed to the significantly reduced minimum Euclidian distance which leads to increased sensitivity to nonlinear effects. However, for a system designed according to single-channel DBP propagation limits, as the one studied here, one can observe that majority of the links operate using EDC alone (except the ones highlighted by up-arrows). Note that managing the PAPR for such formats through linear pre-dispersion could further improve the transmission performance, as shown in Section 1.3. Additionally, in order to examine the available system margin, Figure 9 also shows the received OSNR for various configurations, where it can be seen that majority of the links (except 3) have more than 2 dB available margins, and that our numerical results show an excellent match to the theoretical predictions.

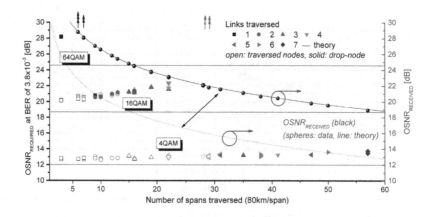

Figure 9. Nonlinear tolerance of PM-mQAM in a dynamic mesh network after EDC. a) Colored: OSNR at BER of 3.8x10⁻³ vs. Distance (Links traversed: 1(square), 2(circle), 3(up-tri), 4(down-tri), 5(left-tri), 6(right-tri), 7(diamond), horizontal lines (theoretical required OSNR)), open: intermediate nodes, solid: destination nodes. Black: Received OSNR (black spheres), Line (theoretical received OSNR), Dotted Line (theoretical received OSNR with 5 dB margin). Up arrows indicate failed connections (corresponding to drop nodes).

As discussed, the results presented in Figure 9 exclude 9 network connections classified as failed (25% of the total traffic), where the calculated BER was always found to be higher than the 3.8x10⁻³. In order to address the failed routes, we employed single-channel DBP, as shown in [21], on such channels, as shown in Figure 10 (red: simplified, blue: full-precision 40 steps per span).It can be seen that all but one of the links can be restored by using single-channel DBP, with the Q_{eff}increasing by an average of ~1 dB, consistent with the improvements observed for heterogeneous traffic in Section 1.3. The link which continues to give a BER even after after single-channel DBP is operated with the highest order modulation format studied, and its two nearest neighbors are both highly dispersed. Note that even though the maximum node lengths are chosen based on nonlinear transmission employing single-channel DBP, most of the network traffic also abide by the EDC constraints (64QAM: ≥ 1 span, 16QAM: ≥ 6 spans, 4QAM ≥ 24 spans). The failed links have one-to-one correlation with violation of these EDC constraints, allowing for prediction of DBP requirements with a quarter of the total network traffic requiring the implementation of single-channel DBP. Also, note that all but two of the links are operable with less than 15 DBP steps for the whole link.

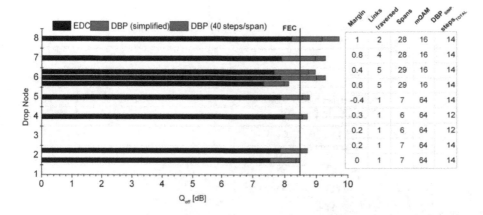

Figure 10. Q_{eff} as a function of network nodes for failed routes, shown by up-arrows in Fig. 5, for PM-mQAM in a dynamic mesh network. After EDC (black) and single-channel DBP (red: simplified; blue: full-precision 40 steps per span). Table shows the network parameters for each scenario and number of steps for single-channel simplified DBP.

These results give some indication of the benefit of flexible formats and DBP. For particular network studied (assuming one of the two failed links works with high precision DBP), if homogeneous traffic, employing 4QAM, is considered, a total network capacity of 4-Tb/s could be achieved. On the other hand, flexible m-ary QAM employing bandwidth allocation based on EDC performance limits only (not shown) enables ~60% increase in transmission capacity (6.8-Tb/s), while designs accounting for SC-DBP add a further 12% increase in capacity (7.7-Tb/s). Note that for traffic calculations based on EDC constraints, we assumed that the routes of Figure 10 would operate satisfactorily for the next format down and that there would be no increase in the nonlinear penalty experienced by any other channel. Further increase in capacity can be attained if pre-dispersion or limited WDM DBP are used, or if more format granularity is introduced (e.g. 8QAM and 32QAM) to exploit the remaining margin. In this example, 25% of transponders operating in single-channel DBP mode enable a 12% increase in capacity. One may therefore argue that in order to provide a the same increase in capacity without employing DBP, approximately 12% more channels would be required, consuming 12% more energy (assuming that the energy consumption is dominated by the transponders). In the case studied, since a ¼ of transponders require DBP, breakeven would occur if the energy consumption of a DBP transponder was 50% greater than a conventional transponder. Given that commercial systems allocate approximately 3-5% of their power to the EDC chipset [32], this suggests that the DBP unit used could be up to 16 times the complexity of the EDC chip. The results reported in Figure 10 with simplified DBP fall within this bound and highlight the practicality of simplified DBP algorithms.

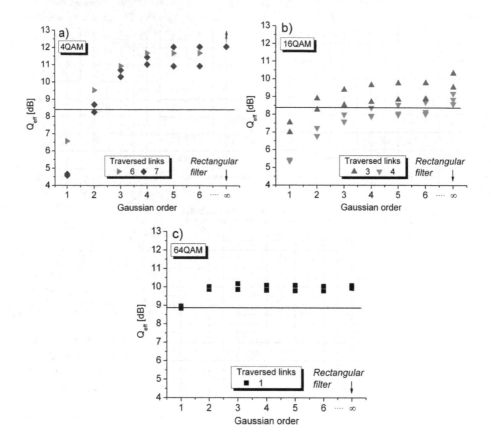

Figure 11. Q_{eff} as a function of Gaussian filter order (35 *GHz* bandwidth) for a 6 *dB* margin from theoretical achievable OSNR. a) 4QAM; b) 16QAM; c) 64QAM. (up-arrows indicate that no errors were detected).

4.2.2. Filter order and BW dependence

Figure 11 shows the performance of a selection of links with less than 6 *dB* margin from the theoretical achievable OSNR (see Figure 9 for links used, we show only the links with the worst required OSNR in the case of 16QAM for clarity), as a function of the Gaussian filter order within each ROADM. As it is well-known, the transmission penalty decreases as filter order increases [33]. However, it can be seen that for higher-order modulation formats, the transmission performance saturates at lower filter orders, compared to lower-order formats. This trend is related to the fact that modulation formats traversing through greater number of nodes are more strongly dependent on the Gaussian order (attributed to known penalties from filter cascades [34,35]). For instance, the performance of 4QAM traffic is severely degraded as a function of Gaussian order, due to the higher number of nodes traversed by such format. 16QAM channels show relatively good tolerance to filter order due to reduced number of hops, however when greater than 3 nodes are employed, the performance again becomes a strong function of filter order. 64QAM is least dependent on filter order since no intermediate ROADMs are traversed. For any given modulation order, a minimum Gaussian order of ~3 enables the optimum performance to be within 1 dB of the performance for an ideal rectangular filter.

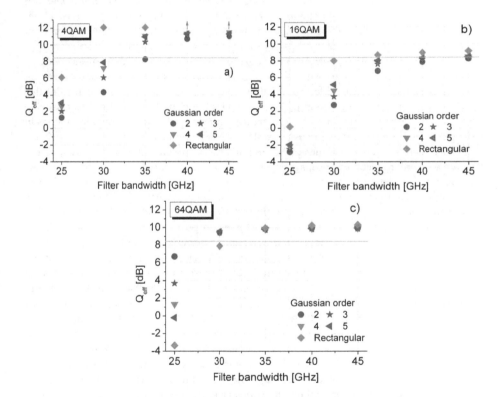

Figure 12. Q_{eff} as a function of Gaussian filter bandwidth (and filter order) for worst-case OSNR margin seen in Figure 6.8. a) 4QAM; b) 16QAM; c) 64QAM.

The simulated Q_{eff} versus 3 dB bandwidth of the ROADM stages and filter order is shown in Figure 12, again for the worst-case required OSNR observed in Figure 9 for each modulation format. For lower bandwidths, the Q_{eff} is degraded due to bandwidth constraints. With the exception of second order filters, bandwidths down to 35 *GHz* are sufficient for all the formats studied. However, consistent with previous analysis (in Figure 10), the impact of filter order on 64QAM is minimal and lower-order filters seem to have better performance than higher-order ones at 25*GHz* bandwidth. This is because when the signal bandwidth (28-*GHz*) exceeds the filter bandwidth, the lower order filters capture more of the signal spectra. However, this effect is visible in the case of 64QAM only since no nodes were traversed in this case, thereby avoiding the penalty from ROADM stages with lower filter orders.

5. Summary and future work

In this chapter we explored the network aspect of advanced physical layer technologies, including multi-level formats employing varying DSP, and solutions were proposed to enhance the capacity of static transport networks. It was demonstrated that that if the order of QAM is adjusted to maximize the capacity of a given route, there may be a significant degradation in the transmission performance of existing traffic for a given dynamic network architecture. Such degradations were shown to be correlated to the accumulated peak-to-average power ratio of the added traffic along a given path, and that management of this ratio through pre-distortion was proposed to reduce the impact of adjusting the constellation size on through traffic. Apart from distance constraints, we also explored limitations in the operational power range of network traffic. The transponders which autonomously select a modulation order and launch power to optimize their own performance were reported to have a severe impact on co-propagating network traffic. A solution was proposed to operate the transponders altruistically, offering lower penalties than network controlled fixed power approach. In the final part of our analysis, the interplay between different higher-ordermodulation channels and the effect of filter shapes and bandwidth of(de)multiplexers on the transmission performance, in a segment of pan-European optical network was explored. It was verified that if the link capacities are assigned assuming that digital back propagation is available, 25% of the network connections fail using electronic dispersion compensation alone. However, majority of such links can indeed be restored by employing single-channel digital back-propagation. Our results indicated some benefit of flexible formats and DBP in realistic mesh networks. We showed that for particular network studied, if homogeneous traffic, employing 4QAM is considered, a total network capacity of 4 *Tb/s* can be achieved. On the other hand, flexible m-ary QAM employing bandwidth allocation based on EDC performance limits enable ~60% increase in transmission capacity (6.8 *Tb/s*), while designs accounting for SC-DBP add a further 12% increase in capacity (7.7 *Tb/s*). Further enhancement in network capacity may be obtained through the use of intermediate modulation order, dispersion pre-compensation for nonlinearity control and the use of altruistic launch powers.

In terms of network evolution, the ultimate goal is to enable software-defined transceivers, where each node would switch itself to *just-right* modulation scheme and associated DSP, based on various physical layer, distance, power, and etc. constraints. Modeling of real-time traffic employing the content covered in this chapter, should motivate and pave the way for high capacity upgrade of currently deployed networks. In addition, modulation/DSP aware routing and wavelength assignment algorithms (e.g. DBP bandwidth aware wavelength allocation) would further enhance the transmission capacity.

Acknowledgements

This work was supported by Science Foundation Ireland under Grant numbers 06/IN/I969 and 08/CE/11523.

Author details

Danish Rafique[1,2] and Andrew D. Ellis[1]

1 Photonic Systems Group, Tyndall National Institute and Department of EE/Physics, University College Cork, Dyke Parade, Ireland

2 now with Nokia Siemens Networks, S.A., Lisbon, Portugal

References

[1] R.W. Tkach, "Scaling optical communications for the next decade and beyond," Bell Labs Technical Journal 14, 3-9 (2010).

[2] P. Winzer, "Beyond 100G Ethernet," IEEE Communications Magazine 48, 26 (2010).

[3] S. Makovejsm, D. S. Millar, V. Mikhailov, G. Gavioli, R. I. Killey, S. J. Savory, and P. Bayvel, "Experimental Investigation of PDMQAM16 Transmission at 112 Gbit/s over 2400 km," OFC/NFOEC, OMJ6 (2010).

[4] J. Yu, X. Zhou, Y. Huang, S. Gupta, M. Huang, T. Wang, and P. Magill, "112.8-Gb/s PM-RZ 64QAM Optical Signal Generation and Transmission on a 12.5GHz WDM Grid," OFC/NFOEC, OThM1 (2010).

[5] M. Seimetz, Higher-order modulation for optical fiber transmission. Springer (2009).

[6] A. Nag, M. Tornatore, and B. Mukherjee, "Optical network design with mixed line rates and multiple modulation formats," Journal of Lightwave Technology 28, 466–475 (2010).

[7] C. Meusburger, D. A. Schupke, and A. Lord, "Optimizing the migration of channels with higher bitrates," Journal of Lightwave Technology 28, 608–615 (2010).

[8] M. Suzuki, I. Morita, N. Edagawa, S. Yamamoto, H. Taga, and S. Akiba, "Reduction of Gordon-Haus timing jitter by periodic dispersion compensation in soliton transmission," Electronics Letters 31, 2027-2029 (1995).

[9] C. Fürst, C. Scheerer, G. Mohs, J-P. Elbers, and C. Glingener, "Influence of the dispersion map on limitations due to cross-phase modulation in WDM multispan transmission systems," Optical Fiber Communication Conference, OFC '01, MF4 (2001).

[10] D.D. Marcenac, D. Nesset, A. E. Kelly, M. Brierley, A. D. Ellis, D. G. Moodie, and C. W. Ford, "40 Gbit/s transmission over 406 km of NDSF using mid-span spectral inversion by four-wave-mixing in a 2 mm long semiconductor optical amplifier," Electronics Letters 33, 879 (1997).

[11] M. Kuschnerov, F. N. Hauske, K. Piyawanno, B. Spinnler, M. S. Alfiad, A. Napoli, and B. Lankl, "DSP for coherent single-carrier receivers," Journal of Lightwave Technology 27, 3614-3622 (2009).

[12] X. Li, X. Chen, G. Goldfarb, Eduardo Mateo, I. Kim, F. Yaman, and G. Li, "Electronic post-compensation of WDM transmission impairments using coherent detection and digital signal processing," Opt. Express, 16, 880 (2008).

[13] D. Rafique, J. Zhao, and A. D. Ellis, "Digital back-propagation for spectrally efficient WDM 112 Gbit/s PM m-ary QAM transmission," Opt. Express 19, 5219-5224 (2011).

[14] C. Weber, C.-A. Bunge, and K. Petermann, "Fiber nonlinearities in systems using electronic predistortion of dispersion at 10 and 40 Gbit/s," Journal of Lightwave Technology 27, 3654-3661 (2009).

[15] G. Goldfarb, M.G. Taylor, and G. Li, "Experimental demonstration of fiber impairment compensation using the split step infinite impulse response method," IEEE LEOS, ME3.1 (2008).

[16] D. Rafique and A. D. Ellis, "Impact of signal-ASE four-wave mixing on the effectiveness of digital back-propagation in 112 Gb/s PM-QPSK systems," Opt. Express 19, 3449-3454 (2011).

[17] L.B. Du, and A. J. Lowery, "Improved single channel backpropagation for intra-channel fiber nonlinearity compensation in long-haul optical communication systems," Opt. Express 18, 17075-17088 (2010).

[18] L. Lei, Z. Tao, L. Dou, W. Yan, S. Oda, T. Tanimura, T. Hoshida, and J. C. Rasmussen, "Implementation Efficient Nonlinear Equalizer Based on Correlated Digital Backpropagation," OFC/NFOEC, OWW3 (2011).

[19] S. J. Savory, G. Gavioli, E. Torrengo, and P. Poggiolini, "Impact of Interchannel Nonlinearities on a Split-Step Intrachannel Nonlinear Equalizer," Photonics Technology Letters, IEEE 22, 673-675 (2010).

[20] D. Rafique and A. D. Ellis, "Nonlinear Penalties in Dynamic Optical Networks Employing Autonomous Transponders," Photonics Technology Letters, IEEE 23, 1213-1215 (2011).

[21] D. Rafique, M. Mussolin, M. Forzati, J. Martensson, M.N. Chugtai, A.D. Ellis, "Compensation of intra-channel nonlinear fibre impairments using simplified digital back-propagation algorithm", Optics Express, Opt. Express 19, 9453-9460 (2011).

[22] C. S. Fludger, T. Duthel, D. vanden Borne, C. Schulien, E.-D. Schmidt, T. Wuth, J. Geyer, E. DeMan, G.-D. Khoe, and H. de Waardt,, "Coherent Equalization and POL-MUX-RZ-DQPSK for Robust 100-GE Transmission," J. Lightwave Technol. 26, 64-72 (2008).

[23] T. Wuth, M. W. Chbat, and V. F. Kamalov, "Multi-rate (100G/40G/10G) Transport over deployed optical networks," OFC/NFOEC, NTuB3 (2008).

[24] W. Wei, Z. Lei, and Q. Dayou, "Wavelength-based sub-carrier multiplexing and grooming for optical networks bandwidth virtualization," OFC/NFOEC, PDP35 (2008)

[25] R. Peter, and C. Brandon, "Evolution to colorless and directionless ROADM architectures," OFC/NFOEC, NWE2 (2008).

[26] D. Rafique and A.D. Ellis "Nonlinear penalties in long-haul optical networks employing dynamic transponders," Optics Express 19, 9044-9049, (2011).

[27] S. Turitsyn, M. Sorokina, and S. Derevyanko, "Dispersion-dominated nonlinear fiber-optic channel," OpticsLetters 37, 2931-2933 (2012) .

[28] L.E. Nelson, A. H. Gnauck, R. I. Jopson, and A. R. Chraplyvy, "Cross-phase modulation resonances in wavelength-division-multiplexed lightwave transmission," ECOC, 309–310 (1998).

[29] D. Rafique, J. Zhao, and A. D. Ellis, "Digital back-propagation for spectrally efficient WDM 112 Gbit/s PM m-ary QAM transmission," Opt. Express 19, 5219-5224 (2011).

[30] D. Rafique and A.D. Ellis "Nonlinear and ROADM induced penalties in 28 Gbaud dynamic optical mesh networks employing electronic signal processing," Optics Express 19, 16739-16748, (2011).

[31] D. Rafique and A. D. Ellis, "Various Nonlinearity Mitigation Techniques Employing Optical and Electronic Approaches," Photonics Technology Letters, IEEE 23, 1838-1840 (2011).

[32] K. Roberts, "Digital signal processing for coherent optical communications: current state of the art and future challenges," SPPCOM, SPWC1 (2011).

[33] F. Heismann, "System requirements for WSS filter shape in cascaded ROADM networks," OFC/NFOEC, OThR1 (2010).

[34] T. Otani, N. Antoniades, I. Roudas, and T. E. Stern, "Cascadability of passband-flat-
tened arrayed waveguidegrating filters in WDM optical networks," Photonics Technology Letters 11, 1414-1416 (1999).

[35] M. Filer, and S. Tibuleac, "DWDM transmission at 10Gb/s and 40Gb/s using 25GHz
grid and flexible-bandwidth ROADM," OFC/NFOEC, NThB3 (2011).

Chapter 2

Optimal Design of a Multi-Layer Network an IP/MPLS Over DWDM Application Case

Claudio Risso, Franco Robledo and Pablo Sartor

Additional information is available at the end of the chapter

1. Introduction

Some decades ago the increasing importance of the telephony service pushed most telecommunications companies (TELCOs) to deploy optical fibre networks. In order to guarantee appropriate service availability, these networks were designed in such a way that several independent paths were available between each pair of nodes, and in order to optimize these large capital investments several models and algorithms were developed.

Already the optimal design of a single layer network is a challenging task that has been considered by many research groups, see for instance the references: [1-3]. Throughout this work this optical network is referred to as the *physical layer*.

Some years afterwards, the exponential growth of Internet traffic volume demanded for higher capacity networks. This demand led to the deployment of dense wavelength division multiplexing (DWDM) technology. Today, DWDM has turned out to be the dominant network technology in high-capacity optical backbone networks. Repeaters and amplifiers must be placed at regular intervals for compensating the loss in optical power while the signal travels along the fibre; hence the cost of a lighpath is proportional to its length over the physical layer. DWDM supports a set of standard high-capacity interfaces (e.g. 1, 2.5, 10 or 40 Gbps). The cost of a connection also depends of the capacity but not proportionally. For economies of scale reasons, the higher the bit-rate the lower the per-bandwidth-cost. The client nodes together with these lightpath connections form a so-called *logical layer* on top of the physical one.

The increasing number of per-physical-link connections -intrinsic to DWDM- may cause multiple logical link failures from a single physical link failure (e.g., fibre cut). This issue led to the development of new multi-layer models aware of the stack of network layers. Most of these models share in common the 1+1 protection mechanism, that is: for every demand two independent lightpaths must be routed such that in case of any single physical link -or even node- failure, at least one of them survive. The following references: [4] and [5] are good examples of this kind of models. Those multi-layer models are suitable for certain families of logical layer technologies such as: synchronous optical networking (SONET) or synchronous digital hierarchy (SDH) since both standards have 1+1 protection as their native protection mechanism.

During many years the connections of IP networks were implemented over SONET/SDH -for simplicity we will only mention SDH from now on-. Most recently: multiprotocol label switching (MPLS), traffic engineering extensions for dynamic routing protocols (e.g. OSPF-TE, ISIS-TE), fast-reroute algorithms (FRR) and other new features were added to the traditional IP routers. This new *technology bundle* known as IP/MPLS, opens a competitive alternative against traditional protection mechanisms based on SDH.

Since IP/MPLS allows recovering from a failure in about 50ms, capital savings may come from the elimination of the intermediate SDH layer. Another improvement of this technology is that the number of paths to route demands between nodes is not pre-bounded; so it might exist in fact a feasible different configuration for most failure scenarios. Since IP/MPLS allows the elimination of an intermediate layer, manages Internet traffic natively, and makes possible a much easier and cheaper operation for virtual private network (VPN) services, it is gaining relative importance every day.

Setting aside technical details and for the purpose of the model presented in this work, we remark two important differences between SDH and IP/MPLS. The first one is the need of SDH to keep different demands between the same nodes. In IP/MPLS networks, all the traffic from one node to another follows the same path in the network referred to as *IP/MPLS tunnel*. The second remarkable difference is how these technologies handle the existence of parallel links in the logical layer. In SDH the existence of parallel links is typical but in IP/MPLS parallel links may conflict with some applications so we will avoid them.

In this paper we address the problem of finding the optimal -minimum cost- configuration of a logical topology over a fixed physical layer. The input data set is constituted by: the physical layer topology -DWDM network-, the client nodes of the logical layer -IP/MPLS nodes- and the potential links between them, as well as the traffic demand to satisfy between each pair of nodes and the per-distance-cost in the physical network associated with the bitrates of the lightpaths to deploy over it. The decision variables are: what logical links do we have to implement, which bitrate must be assigned to each of them and what path do these lightpaths have to follow in the physical layer. For being a feasible solution a configuration must be capable of routing every traffic demand over the remaining active links of the logical layer for every single physical link failure scenario.

The problem previously described is NP-hard and due to its complexity we developed a metaheuristic based on GRASP to find good quality solutions for real size scenarios. Actually, we analysed the performance of the proposed metaheuristic using real-world scenarios provided by the Uruguayan national telecommunications company (ANTEL).

The main contributions of this article are: i) a model to represent a common network overlay design problem; ii) the design of a GRASP metaheuristic to find good quality solutions for this model; iii) the experimental evaluation based on real-world network scenarios.

The remaining of this document is organized as follows. A mixed-integer programming model will be presented in Section-2. In Section-3 we will show some exact solutions found with CPLEX for small/simple but illustrative problems; in this section we also analyze the intrinsic complexity of the problem. In Section-4 a GRASP metaheuristic to solve this problem is presented. Finally, in Section-5 we will show the solutions found with the previous metaheuristic for real-world network scenarios.

2. Mathematical model

We will now introduce the basic mixed-integer programming model that arises from the detailed interaction of technologies.

2.1. Parameters

The physical network is represented by an undirected graph (V, P), and the logical network is represented by another undirected graph (V, L). Both layers share the same set of nodes. The links of the logical layer are potential -admissible logical links- while the links of the physical layer are definite. In both graphs the edges are simple since multigraphs are not allowed in this model.

For every different pair of nodes $p, q \in V$ is known the traffic volume d_{pq} to fulfil along the unique path (tunnel) this traffic follows throughout a logical layer configuration.

These paths are unique at every moment, but in case of link failures they may change to follow an alternate route. For simplicity we assume that the traffic volume is symmetric (i.e. $d_{pq} = d_{qp}$). Let $B = \{b_1, \ldots, b_{\bar{B}}\}$ be the set of possible bitrate capacities for the lightpaths on the physical layer and therefore for the links of the logical one. Every capacity $b \in B$ has a known per-distance cost c_b. For economies of scale reasons it holds that if $b' < b''$ then $(c_{b'}/b') > (c_{b''}/b'')$. Since both graphs of this model are simple and undirected, we will express links as pairs of nodes. For every physical link (ij) is known its length l_{ij}.

2.2. Variables

This model comprises three classes of variables. The first class is composed of the logical link capacity variables. We will use boolean variables τ_{pq}^b to indicate whether or not the logical link $(pq) \in L$ has been assigned with the capacity $b \in B$. As a consequence the capacity of the link (pq) could be computed as: $\sum_{b \in \hat{B}} b \cdot \tau_{pq}^b$.

The second class of variables determines how are going to be routed the logical links over the physical network. If $\sum_{b \in \hat{B}} \tau_{pq}^b = 1$ then the logical link $(pq) \in L$ was assigned with a capacity, it is going to be used in the logical network and requires a lightpath in the physical one. y_{pq}^{ij} is a boolean variable that indicates whether or not the physical link $(ij) \in P$ is being used to implement the lightpath of $(pq) \in L$. Since lightpaths cannot automatically recover from a link failure, whenever a physical link (ij) fails all the logical links (pq) such that $y_{pq}^{ij}=1$ do fail as well. The only protection available in this model is that of the logical layer. For demands being protected against single physical link failures, it is necessary to have a feasible route through the remaining active logical links.

The third and final class of variables is that that determines how the IP/MPLS tunnels are going to be routed against any particular failure in the physical layer. $^{rs}x_{pq}^{ij}$ is a boolean variable that indicates whether the logical link $(pq) \in L$ is going to be used or not, to route traffic demand $d_{rs}>0$, under a fault condition in the physical link $(ij) \in P$.

NOTE: To keep the nomenclature of the variables as easy as possible we always placed: logical links subindexes at bottom right position, physical links subindexes at top right position and demands subindexes at top left position.

2.3. Constraints

This problem comprises three groups of constraints. The first group of constraints establishes the rules that the routes of the lightpaths must follow to be feasible.

$$\sum_{b \in B} \tau_{pq}^{b} \leq 1 \qquad \forall (pq) \in L. \tag{1}$$

$$\sum_{j/(pj) \in P} y_{pq}^{pj} = \sum_{b \in B} \tau_{pq}^{b} \qquad \forall (pq) \in L. \tag{2}$$

$$\sum_{i/(iq) \in P} y_{pq}^{iq} = \sum_{b \in B} \tau_{pq}^{b} \qquad \forall (pq) \in L. \tag{3}$$

$$\sum_{j/(ij) \in P} y_{pq}^{ij} = 2\hat{\theta}_{pq}^{i} \qquad \substack{\forall (pq) \in L, \forall i \in V, \\ i \neq p, i \neq q.} \tag{4}$$

$$\tilde{\theta}_{pq}^{i} + \hat{\theta}_{pq}^{i} = 1 \qquad \substack{\forall (pq) \in L, \forall i \in V, \\ i \neq p, i \neq q.} \tag{5}$$

$$y_{pq}^{ij} - y_{pq}^{ji} = 0 \qquad \forall (pq) \in L, \forall (ij) \in P. \tag{6}$$

$$\forall (pq) \in L, \forall (ij) \in P. \qquad \begin{array}{l} \forall (pq) \in L, \forall (ij) \in P, \\ \forall b \in \hat{B}, \forall i \in V. \end{array} \tag{7}$$

The meaning of the previous constraints is the following: (1) establishes that the number of capacities assigned to every logical link is at most 1 -it could be 0 if the link is not going to be used-; (2) and (3) guarantee that if any particular link $(pq) \in L$ was assigned with a capacity $(\sum_{b \in \hat{B}} \tau_{pq}^{b} = 1)$ then there must exist one and only one outgoing -or incoming- physical link used for its lightpath.

Before going any further we have to introduce a set of auxiliary variables: θ_{pq}^{i} and $\hat{\theta}_{pq}^{i}$. These variables are defined for every combination of logical links $(pq) \in L$ and physical nodes $i \in V$. (5) guarantees that exactly one of the following conditions must meet: $(\theta_{pq}^{i}, \hat{\theta}_{pq}^{i}) = (1, 0)$ or $(\theta_{pq}^{i}, \hat{\theta}_{pq}^{i}) = (0, 1)$. Hence, (4) guarantees flow balance for routing the lightpaths through the remaining -not terminal- nodes. Finally (6) guarantees that the lightpaths go back and forth through the same path, while (7) stands the integrity of the variables.

The second group of constraints establishes the rules that the routes of the IP/MPLS tunnels must follow in the logical layer and their meaning is similar to the previous ones except for (1) and (8). The inequalities in (8) were added to guarantee that whatever the failure scenario is ($\forall (ij) \in P$), its associated routing configuration over the logical network keeps the aggregated traffic load below the link capacity for every data link ($\forall (pq) \in L$). Constrains (2) and (3) are equivalent to (9) and (10), except for the fact that in the latter the existence of a tunnel relies on the existence of demand and this is known in advance. Another remarkable point is that the second group of constraints has as many possible routing scenarios as arcs in P, so the number of variables is much greater.

$$\sum_{rs:d_{rs}>0} d_{rs} \cdot {}^{rs}x_{pq}^{ij} \leq \sum_{b \in \hat{B}} b \cdot \tau_{pq}^{b} \quad \forall (pq) \in L, \forall (ij) \in P. \tag{8}$$

$$\sum_{q/(rq) \in L} {}^{rs}x_{rq}^{ij} = 1 \qquad \forall d_{rs}>0, \forall (ij) \in P. \tag{9}$$

$$\sum_{p/(ps) \in L} {}^{rs}x_{ps}^{ij} = 1 \qquad \forall d_{rs}>0, \forall (ij) \in P. \tag{10}$$

$$\sum_{q/(pq) \in L} {}^{rs}x_{pq}^{ij} = 2 \cdot {}^{rs}\hat{\mu}_{p}^{ij} \qquad \begin{array}{l} \forall d_{rs}>0, \forall (ij) \in P, \\ \forall p \in V, p \neq r, p \neq s. \end{array} \tag{11}$$

$$ {}^{rs}\tilde{\mu}_{p}^{ij} + {}^{rs}\hat{\mu}_{p}^{ij} = 1 \qquad \begin{array}{l} \forall d_{rs}>0, \forall (ij) \in P, \\ \forall p \in V, p \neq r, p \neq s. \end{array} \tag{12}$$

$$^{rs}x_{pq}^{ij} - {}^{rs}x_{qp}^{ij} = 0 \qquad \begin{array}{l} \forall d_{rs}>0, \forall (pq)\in L, \\ \forall (ij)\in P. \end{array} \tag{13}$$

$$^{rs}x_{pq}^{ij}, {}^{rs}\tilde{\mu}_p^{ij}, {}^{rs}\hat{\mu}_p^{ij} \in \{0,1\} \qquad \begin{array}{l} \forall d_{rs}>0, \forall (pq)\in L, \\ \forall (ij)\in P, \forall p\in V. \end{array} \tag{14}$$

Variables sets $^{rs}\tilde{\mu}_{pq}^i$ and $^{rs}\hat{\mu}_{pq}^i$ are homologous to θ_{pq}^i and $\hat{\theta}_{pq}^i$; so are constraints from (4) to (7) with those from (11) to (14). Before proceeding any further we must notice that both groups are not independent. Many logical links may not be available for routing after a physical link failure. Which logical links are in this condition, relies on how the lightpaths were routed in the physical layer. Specifically, if some logical link (pq) uses a physical link (ij) for its lightpath implementation then this logical link cannot be used to route any tunnel under (ij) failure scenario.

$$\left\{ {}^{rs}x_{pq}^{ij} \leq 1 - y_{pq}^{ij} \qquad \forall rs : d_{rs} > 0, \forall (pq) \in L, \forall (ij) \in P, \right. \tag{15}$$

The group of constrains (15) prevents from using (pq) to route any traffic ($^{rs}x_{pq}^{ij}=0$, $\forall rs : d_{rs}>0$) in any failure scenario which affects the link (when $y_{pq}^{ij}=1$).

2.4. Objective

The function to minimize is the sum of the cost of every logical link. According on what capacity was assigned to a logical link there is an associated per-distance-cost (c_b), and according on how the corresponding lightpath was routed over the physical layer there is an associated length ($\sum_{(ij)\in P} l_{ij} y_{pq}^{ij}$). The product of both terms is the cost of a particular logical link and the sum of these products for all of the logical links is the total cost of the solution. The direct arithmetic expression for the previous statement would be: $\sum_{(pq)\in L} (\sum_{b\in\hat{B}} c_b \tau_{pq}^b)(\sum_{(ij)\in P} l_{ij} y_{pq}^{ij}) = \sum_{(pq)\in L, (ij)\in P, b\in\hat{B}} c_b l_{ij} \cdot \tau_{pq}^b y_{pq}^{ij}$.

Although straightforward, this approximation is inappropriate because it is not linear. The following sub-problem expresses the objective value with an equivalent linear expression.

$$\min \sum_{\substack{(pq)\in L \\ (ij)\in P \\ b\in B}} c_b l_{ij} \cdot {}^b\eta_{pq}^{ij} \tag{16}$$

$$^b\eta_{pq}^{ij} \geq \tau_{pq}^b + y_{pq}^{ij} - 1 \qquad \begin{array}{l} \forall (pq)\in L, \forall (ij)\in P, \\ \forall b\in\hat{B}. \end{array} \tag{17}$$

$$b_\eta{}^{ij}_{pq} \geq 0 \qquad \begin{array}{l}\forall(pq)\in L,\forall(ij)\in P,\\ \forall b\in \hat{B}.\end{array} \tag{18}$$

We used the real variable ${}^b\eta{}^{ij}_{pq}$ instead of $\tau^b_{pq} y^{ij}_{pq}$ and added some extra constraints to guarantee the consistency. This consistency comes from the following observations: the result of $\tau^b_{pq} y^{ij}_{pq}$ is also a boolean variable, and since ${}^b\eta{}^{ij}_{pq}$ is being multiplied by a positive constant in a minimization problem it will take its lowest value whenever this is possible. This value would be zero because of constrains (3) of (D).

The only exception is when the values of τ^b_{pq} and y^{ij}_{pq} are both 1, in which case the value of ${}^b\eta{}^{ij}_{pq}$ should be 1 as well to keep consistency. This is guaranteed by constrain (17). The complete MIP is the result of merging (1) to (18).

3. Finding exact solutions

We will start by showing particular solutions for some simple example cases. The first example has four nodes $V = \{v_1, v_2, v_3, v_4\}$, the physical layer is the cycle (C^4) while the logical layer is the clique (K^4). The remaining parameters are: $\hat{B}=\{3\}$, $d_{pq}=1$ $\forall$ $1\leq p<q\leq4$ and $l_{ij}=1$, $\forall$ $(ij)\in P$. c_b is irrelevant in this case because there is only one bitrate available. The optimal solution found for this case uses all of the logical links. Figure 1 shows with dashed lines the route in that solution followed for each lightpath over the physical cycle. This is an example where lightpaths routes are not intuitive, even for a very simple input data set.

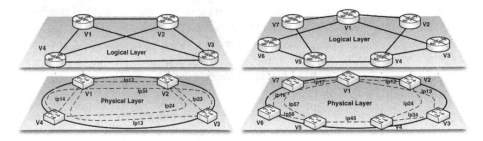

Figure 1. Optimal solutions found for: K^4 over C^4, with $d_{pq}=1$ and $\hat{B}=\{3\}$, and for: K^7 over C^7, with $d_{1q}=1$ and $\hat{B}=\{3\}$.

The following example comprises seven nodes and explores again the clique-over-cycle case. The remaining parameters are analogous: $B=\{3\}$, $l_{ij}=1$ $\forall$ $(ij)\in P$, except for demands that in this case are to/from one single node ($d_{1q}=1, \forall$ $1<q\leq7$). Unlike the previous example, the optimal solution in this case (also sketched in Figure 1) does not make use of all the logical links. Although the route followed by each lightpath looks more natural in this example, it is

not immediate why this set of logical links ought to be the appropriate to construct the optimal solution.

Through these two examples we attempted to show that solutions are not intuitive even for very simple cases. To find optimal solutions we used ILOG CPLEX v12.1. All computations were performed on a Linux machine with an INTEL CORE i3 Processor and 4GB of DDR3 RAM. Table 1 shows information for several test instances analogous to those represented in Figure 1, that is: K^n over C^n with $d_{pq}=1$, $\forall\, 1 \leq p < q \leq n$ and $l_{ij}=1$, $\forall\, (ij) \in P$ over a range of integer b_1 values ($|\hat{B}|=1$).

$|V|$	b_1 range	#variables	#constrains	Elapsed time (hh:mm:ss)
5	2 – 6	1230	1640	00:00:00 – 000:00:11
6	3 – 9	3390	4035	00:00:02 – 000:19:31
7	2 – 12	7896	8652	(*) 00:00:05 – 087:19:05
8	3 – 16	16296	16772	(*) 00:00:02 – 100:10:17

(*)Note: The solver aborted for some intermediate cases

Table 1. Overall results for some particular cases

We proved that: it is always possible to find minimal feasible solutions for these particular topologies and demands when: $b_1=2$ and $|V|$ is odd, or when $b_1=3$ and $|V|$ is even. In the first situation the complete logical graph is needed, whereas in the second only diagonal links can be disposed of. The lowest computation times were found for these extreme cases.

We also proved that: the cycle configuration for the logical network -the simplest possible- is feasible for every b_1 greater or equal to: $|V|^2/4$ when $|V|$ is even, or $(|V|^2-1)/4$ when $|V|$ is odd. Very low computation times were found for these cases also. The time required for finding optimal solutions for non-extreme cases were much higher. CPLEX even aborted for many of them. Aside from a bunch of worthless exceptions, we couldn't find solutions for topologies other than K^n over C^n.

Keeping these physical and logical topologies while trying with simpler matrices of demand (e.g. $d_{1q}=1, \forall\, 1 < q \leq n$) it was possible to increase the size of the problems to 15 nodes and yet being able to find optimal solutions. Suffices to say this size bound as well as the simplicity in the topologies and traffic matrices of the previous examples, are incompatible with real life network problems.

Proposition 1: The problem presented in this section is NP-Hard.

Demonstration lies under reduction of NPP (Number Partitioning Problem) to our particular problem that will be referred to as MORNP (Multi-Overlay Robust Network Problem) within this proof. NPP problem consist in finding two subsets with the same sum for a known multiset of numbers. Formally: given a list of positive integers: $a_1, a_2, \ldots, a_N$, a partition $A \subseteq \{1, 2, \ldots, N\}$ must be found so that discrepancy:

$$E(A) = \left| \sum_{i \in A} a_i - \sum_{i \notin A} a_i \right|, \tag{19}$$

finds its minimum value within the set $\{0, 1\}$. NPP is a very well known NP-Complete problem (see for instance [6]).

($\Rightarrow$) Given such a list of positive integers we create an instance of MORNP by taking: $B = \{b_1 = \lceil (\sum_{1 \leq i \leq N} a_i)/2 \rceil\}$, logical and physical graphs with the same topology schematized in Figure 2, $l_{ij} = 1 \ \forall \ (ij) \in P$ and $d_{iD} = a_i, \forall \ 1 \leq i \leq N$.

Since logical and physical topologies are the same and all distances are equal, the logical layer projected over the physical one for any optimal solution must copy the underlying shape. So, if there exists a solution for such an instance of MORNP, this solution should have a feasible routing scenario when transport -and logical- link $(v_{A1}v_{A2})$ fails and therefore a way to accommodate traffic requirements over $(v_{H1}v_D)$ and $(v_{H2}v_D)$, due to the fact that both links are still in operational state and they are the only way to reach v_D.

Because there is only one capacity both links must have been assigned with b_1, this can only be done when discrepancy is not grater than one, so we indirectly found a solution for the original NPP problem.

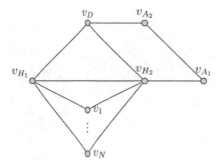

Figure 2. Graph used for NPP reduction to MORNP.

($\Leftarrow$) The complementary part of the proof is easier. Given a solution to an instance of NPP, this partition is used to distribute tunnels between $(v_{H1}v_D)$ and $(v_{H2}v_D)$. Once in v_{H1} or v_{H2} the tunnel is terminated directly in the corresponding node, except for some fault condition in one of these links, in which case a detour through the other v_{Hx} node is always possible. When the fault condition arises in $(v_{H1}v_D)$ or $(v_{H2}v_D)$, a detour may be taken through: $(v_D v_{A2})$, $(v_{A2}v_{A1})$, $(v_{A1}v_{H2})$ or $(v_D v_{A2})$, $(v_{A2}v_{A1})$, $(v_{A1}v_{H2})$, $(v_{H2}v_{H1})$.

Since all the transformation are of polynomial complexity it stands that $NPP \prec MORNP$ and MORNP is NP-Hard.

We proved the complexity of MORNP is intrinsic to the problem, since it is NP-Hard. Because of the previous result and like for most other network design problems, an exhaustive search for the optimal solution of the problem presented in this work is infeasible for real size problems.

4. Metaheuristics

We decided to use a metaheuristic algorithm based on GRASP to find good quality solutions for real instances of this problem. A very high level diagram of our algorithm is shown in Figure 3.

4.1. GRASP implementation

As for every GRASP implementation this algorithm has a loop with two phases. The *construction phase* builds a *randomized feasible solution*, from which a local minimum is found during the *local search phase*. This procedure is repeated *MaxIter* times, while the best overall solution is kept as the result. Further information and details in GRASP algorithms can be found in [7] or in [8].

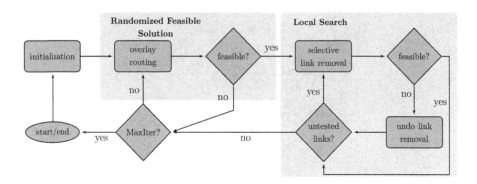

Figure 3. Block-diagram of the GRASP implementation used.

The *initialization phase* performs computations whose results are invariants among the iterations, like the shortest path and distance over the physical layer between each pair of nodes.

4.2. Construction phase

The *randomized feasible solution phase* performs a heuristic low cost balanced routing of the logical layer over the physical one. The exact solution for this sub-problem is also NP-Complete as it can be seen in [9]. The goal is to find a path for every lightpath, such that the number of physical link intersections is minimum. It is also desirable that the total cost be as low as possible but as a secondary priority. The strategy chosen in this heuristic is the following: nodes are taken randomly (.e.g.: uniformly), and for each node their logical links are also taken randomly but with probabilities in inverse ratio to the minimal possible distance of their lightpaths over the physical layer.

Instead of using the real distances of the physical links (l_{ij}), from this point on and until the next iteration pseudo-distances: l_{ij}, $\forall$ $(ij) \in P$ will be used. Prior to start routing lightpaths, all these pseudo-distances are set to 1.

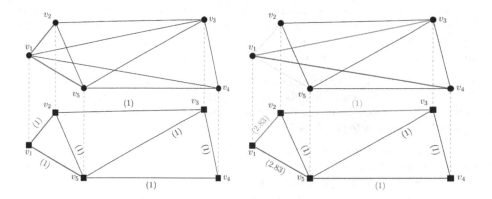

Figure 4. An example of the balanced routing heuristic.

According to these new weights, logical links are routed following the minimal distance without repeating physical links among them. Usually, after routing some lightpaths the set of not-yet-used physical links empties, and it is necessary to start over a new *control window* by filling again the not-yet-used set. Prior to do this, the pseudo-distances are updated using the following rule: $l_{ij} = (1 + n_{ij})^p$ for some fixed penalty p, where n_{ij} is the number of lightpaths that are making use of $(ij) \in P$ up to the moment.

For instance, let us guess our networks are like those sketched in Figure 4 and the links drawn are: (12), (15), (13), (14), (23), (35), and so on. The left half of Figure 4 shows with red and blue lines how are routed the lightpaths (12) and (15). At this point we need to update the pseudo-distances and restart the window. If $p=1.5$ and since $n_{12}=n_{15}=1$, then $l_{12}=l_{15}=2^{1.5}\approx 2.83$ for the next window.

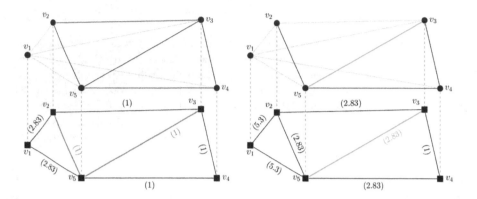

Figure 5. Lightpaths for logical links (23) and (35).

The next two logical links are (13) and (14). They are routed using the updated values. Their lightpaths are also represented with green and magenta lines in the right half of Figure 4. The link (23) is the following and it can be routed in two hops. A window restart is necessary to route the lightpath of (35), as it can be seen in Figure 5.

The elements of the input data in Algorithm-1 are: the logical graph (V, L), the physical graph (V, P), the minimum distance over the physical layer to connect each pair of nodes -computed in the *initialization phase*-. The output is an application between logical links and the subset of physical links used by their lightpaths.

The algorithm detailed in Algorithm-1 is the one depicted in Figure 4 and Figure 5. The outcome of the randomized feasible solution phase is a candidate configuration for the route of each lightpath over the physical network. We did not make use of capacity and traffic information yet; and before going any further we must state that -as in the exact examples- in our practical applications we limited the capacities set to only one capacity ($|\hat{B}| = 1$).

The main reason was that the telecommunications company we developed this application for, wanted the maximum possible bitrate for all the interfaces of its core network. The next issue is determining whether the configuration found is feasible or not. The answer to this question is far from being easy, since this sub-problem is NP-Complete. We have based on a heuristic to answer this question. The heuristic is the following: demands are taken in decreasing order of volume (d_{pq}) and each tunnel is routed over the logical layer following the minimal number of hops, but using only links with remaining capacity to allocate the new tunnel demand.

For instance, Figure 6 shows an example logical topology whose link capacities are 3. Let the demands be: $d_{24}=2$, $d_{12}=1$, $d_{13}=1$ and $d_{23}=1$. The path followed by every tunnel is sketched in Figure 6 using: violet, orange, red, and green curves respectively; so it is the remaining capacity in every link after routing each tunnel -two tunnels in the central image-.

Overlay routing (logical over physical). Algorithm 1

Input: $(V, L), (V, P), d : V \times V \to \mathbb{R}_0^+$. **Output:** $\Psi : L \to 2^P$.

1: Set $\Psi(e) = \emptyset, \forall e \in L$ and $pd : P \to \mathbb{R}_0^+, pd(e) = 1 \, \forall e \in P$.
2: **while** $\exists v \in V$ / not-processed(v) **do**
3:    Select randomly $v \in V$ / not-processed(v));
4:    Set $prob(vw) = \frac{1}{d(v,w)}, \forall (vw) \in L / \Psi(vw) = \emptyset$;
5:    Normalize $prob$ such that: $\sum_{e \in L} prob(e) = 1$;
6:    Create new $control\ window$;
7:    **while** $\exists w \in V / ((vw) \in L$ **AND** $\Psi(vw) = \emptyset)$ **do**
8:       Draw such $w \in V$ randomly weighted by $prob(vw)$;
9:       Find $shortest\text{-}lp$, a $pd\text{-}distance$ shortest lightpath for (vw) avoiding repeating physical
         links within this window;
10:      **if** $(shortest\text{-}lp=\emptyset)$ AND (there are not unprocessed $(vw))$ **then**
11:         Update $pd(v, w) = pd(w, v) = (1 + \sum_{e \in L} |\Psi(e) \cap \{(vw)\}|)^p$;
12:         Create new $control\ window$;
13:      **else**
14:         $\Psi(v_e v_f) = \Psi(v_f v_e) = shortest\text{-}lp$;
15:      **end if**
16:   **end while**
17: **end while**
18: **return** $\Psi : L \to 2^P$.

Algorithm 1.

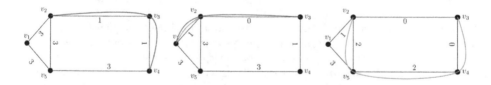

Figure 6. Routes for the tunnels (24), (12), (13), and (23) over a Logical Layer.

This constraint based routing algorithm is straightforward and it is based on Dijkstra's algorithm. Nevertheless an efficient implementation is quite complicated because of the following fact: to be sure a solution is feasible this algorithm must be repeated for each single failure scenario. In order to improve the efficiency: routes cache, optimized data structures and several others low-level programming techniques were used. This isFeasible function is used in both: construction and local search phases. The performance of this function is critic since it is used several times within the same iteration in the local search, as it is represented in Figure 3. Up to this point and before entering the local search phase, we have a feasible configuration for the routes of every lightpath; but we are still using all of the initial logical links and this input network is very likely to be over-sized. Moreover, in the construction phase we attempted to distribute the routes of the lightpaths uniformly over the physical layer, but it is still possible that many logical links fail simultaneously because of a single physical link failure. Therefore, it is very likely that many of these "redundant links" may be disposed of, if they are not really adding useful capacity. It is worth mentioning that from this point on and until the next iteration, lightpaths costs are revealed because we have their lengths -from the configuration for their routes- and there is only one possible capacity.

4.3. Local search

Through the local search phase we intend to remove the most expensive and unnecessary logical links for the current configuration. The process is the following: logical links are taken in decreasing order of cost for their lightpaths, each one is removed and the feasibility of the solution is tested again. If the solution remains feasible the current logical link is permanently removed, otherwise it is reinserted and the sequence follows for the remaining logical links. Once this processes is finished the result is a minimal solution. After *MaxIter* iterations the best solution found is chosen to be the output of the algorithm. Since the construction procedure we have used in this work privileges the nodes drawn earlier to shape the routes of the lightpaths, we presume that adding path-relinking to this algorithm could significantly improve the quality of the result, if the initial lightpaths routes of the elite solutions are prioritized to explore new solutions. We are planning to check this assumption in a future work. For further information in path-relinking enhancement to GRASP, please refer to: [7] and [10].

5. Application case context and results

We will focus now in the context of ANTEL -the telecommunication company we applied this metaheuristic to-. Prior to doing so we are giving some basic elements of the overall Internet architecture. Internet is actually a network that could be disaggregated into several separate smaller networks also known as *Autonomous Systems* (AS). Typically every AS is a portion of the global Internet owned/governed by a particular Internet Service Provider (ISP). Internet users access content residing in servers of: companies, universities, government sites or even from other residential customers (e.g. P2P applications). A portion of this content is located within the own network of the ISP this customer lease the service to -into some of the *Points Of Presence* (POP) of the ISP-, but most content is scattered over the Internet. Since traffic interchange is necessary among different ISPs, the Internet architecture needs special POPs known as *Network Access Points* (NAPs). Within these NAPs: Carriers, ISPs and important content providers (e.g.: Google, Akamai) connect to each other in order to interchange traffic.

This company had two different IP/MPLS networks referred to as: *aggregation network* and *public Internet network*. The *aggregation network* is geographically dispersed all over the country and it is responsible of gathering and delivering the traffic of the customers to the *public Internet network*. The *public Internet network* is where the AS of this ISP is implemented; centralizes the international connections with other ISPs as well as those to Datacenters of local content providers. The *public Internet network* is geographically concentrated and only has POPs in the Capital City and in an important NAP of the US territory (see grey clouds in Figure 7). In terms of the model covered in this article we may stand that the physical network has all of its nodes but one -the NAP- within the national boundaries.

There are four independent paths for international connections -leased to Carriers- between the NAP and the national boundaries. The *aggregation* and *public Internet* networks are both logical. The *public Internet network* only has presence in a few POPs of the Capital City and in the NAP; and although the *aggregation network* has full-national presence it does not span to the NAP. A Non-Disclosure Agreement (NDA) signed between the telecommunications company and our research institute protects more accurate information and details. The costs and traffic information shown in the rest of this article are only referential.

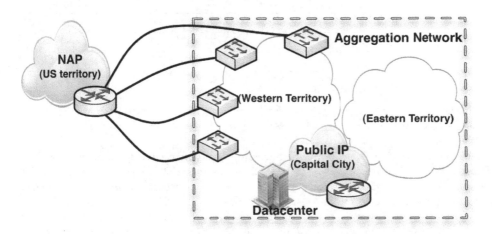

Figure 7. Remarkable aspects of the particular network architecture.

Several planning concerns arose from the situation exposed: Is it convenient the current architecture? Or it would be better to merge both IP/MPLS networks? Are profitable the IT infrastructure investments necessary to increase the percentage of local content? Which would be the optimal network to fulfil every demand requirements scenario? We helped to answer these questions by identifying representative scenarios and creating their associated data sets to feed the metaheuristic.

The overall performance of the algorithm described in Section-4 was very good -under the two hours of execution time in every scenario-.

We tried several scenarios based on the following considerations: traffic volume, network architecture and the percentage of locally terminated traffic. We selected eight remarkable scenarios to detail in Table 2.

According on traffic forecasts it is expected that some years from now the total volume of traffic to be placed somewhere between 57 and 100 (reference values). If some IT investments and agreements were made it is expected that the percentage of locally terminated traffic (national traffic) could be greater (High). These new potential sources of traffic would be placed in the Capital City, specifically in the same POPs where the public Internet network is present. Those scenarios where merged networks is set to False inherit the current network architecture.

scenario index	aggregated traffic demand	%local content	merged networks	number of nodes	required lightpaths	total cost
1	100	Low	False	56	81	10,000,000
2	100	Low	True	68	118	7,662,651
3	100	High	False	56	81	7,578,234
4	100	High	True	68	133	5,713,563
5	57	Low	False	56	75	6,319,470
6	57	Low	True	63	94	4,872,987
7	57	High	False	56	75	5,108,587
8	57	High	True	63	105	4,064,597

Table 2. Referential results for representative scenarios

Another remarkable aspect of this architecture is that whereas the aggregation network is deployed directly over the physical layer, there is an extra SDH layer between the public Internet network and the physical one. Since the public Internet network only has a few nodes and its protection relies on the 1+1 protection mechanism of SDH, its optimal value can be estimated easily. The only portion where we needed computer assistance is that of the aggregation network. The columns number of nodes and required lightpaths refers exclusively to the values for this last network.

On the other hand and in order to compare solutions fairly, the column *total cost* represents the combined cost of both networks -when they are not joined-. It is worth observing those scenarios: 1 and 3, as well as 5 and 7 require the same number of lightpaths. Moreover, their solutions use exactly the same lightpaths. This result should be expected because in both pairs of scenarios share the same traffic and non-merged network architecture; since Datacenters - the only difference- are connected to the *public Internet network*, the *aggregation network* is unaware of the percentage of local content. The only changes are in the *total cost* because of the saving of international capacity.

Less intuitive are those savings arising exclusively from the merging of both networks like: 1 and 2, 3 and 4, and so on. The reason is the following: "the routing search-space of the IP/MPLS technology is much greater than that of the SDH equivalent, so it is much more efficient". For simplicity let us guess for a while that traffic does not need to be fitted in tunnels and instead can behave as a fluid. Since the length of international connections is measured in thousands of kilometres, this links are the most expensive of the physical network. As it was showed in Figure 7 there are four independent connections to the NAP; hence if we needed to guarantee 60Gbps of international traffic we could reserve 20Gbps in every one of these links, because a single failure could only affect one of them. Therefore the efficiency in the usage of international connections could rise to 75% if the efficiency of IP/MPLS would be available. The protection mechanism of SDH (1+1) cannot exploit this degree of connectivity. To protect 60Gbps of traffic using SDH active/stand-by independent paths, we always need other 60Gbps of reserved capacity, so the efficiency of SDH it is limited to 50%. The improved efficiency of IP/MPLS to exploit the extra connectivity degree between local and international traffic explains by itself most of the savings.

6. Conclusion

Perhaps the most remarkable result of this work relies on exposing through a real-world application, how much more cost-efficient could be networks deployed using IP/MPLS, than those based on traditional protection schemes like SDH. This efficiency comes not only from the savings linked to the elimination of an intermediate layer, but also from the extra degrees of freedom available to route the traffic.

We presume that the application this work dealt with is not an exception, and the potential savings might replicate from one ISP to the other. The results for the examples of Section-5 and their later analysis justify the convenience to update network design models this work introduced, in order to follow the new technology trends and exploit their benefits.

Regarding on the metaheuristic, we presume that applying path-relinking could significantly increase the computational efficiency of the proposed GRASP, so this is the line of our immediate future work.

Author details

Claudio Risso*, Franco Robledo and Pablo Sartor*

*Address all correspondence to: crisso@fing.edu.uy

Engineering Faculty – University of the Republic, Montevideo, Uruguay

References

[1] Okamura H. and Seymour P., (1981), "Multicommodity flows in planar graphs", Journal of Combinatorial Theory 31(1), pp. 75–81.

[2] Stoer M., (1992), "Design of survivable networks", Lecture Notes in Mathematics.

[3] Kerivin H. et al., (2005), "Design of survivable networks: A survey", Networks 46(1), pp. 1–21.

[4] Orlowski S. et al., (2007), "Two-layer network design by branch-and-cut featuring MIP-based heuristics", Proceedings of the 3rd International Network Optimization Conference (INOC).

[5] Koster A. et al., (2008), "Single-layer Cuts for Multi-layer Network Design Problems", Proceedings of the 9th INFORMS Telecommunications Conference, Vol. 44, Chap. 1, pp.1-23.

[6] Mertens. S., (2006) "The Easiest Hard Problem: Number Partitioning", "Computational Complexity and Statistical Physics", pp.125-139, Oxford University Press, New York, http://arxiv.org/abs/cond- mat/0302536.

[7] Resende M., Riberio C., (2003), "Greedy randomized adaptive search procedures", ATT Research, http://www2.research.att.com/~mgcr/doc/sgrasp-hmetah.pdf.

[8] Resende M., Pardalos P., (2006), "Handbook of Optimization in Telecommunication", Springer Science + Business Media.

[9] Oellrich M., (2008), "Minimum Cost Disjoint Paths under Arc Dependence", University of Technology Berlin.

[10] Glover F., (1996), "Tabu search and adaptive memory programming - Advances, applications and challenges", Interfaces in Computer Science and Operations Research, pp.1-75.

Chapter 3

Faults and Novel Countermeasures for Optical Fiber Connections in Fiber-To-The-Home Networks

Mitsuru Kihara

Additional information is available at the end of the chapter

1. Introduction

The number of subscribers to broadband services in Japan now exceeds 34 million, and about 20 million were using fiber-to-the-home (FTTH) services in December 2011 [1]. The number of optical fiber cables continues to increase as the number of FTTH subscribers increases; however, unexpected faults have occurred along with this increase. One such fault is damage caused by wildlife including rodents, insects, and birds [2], and another is that caused by defective optical fiber connectors [3]. It is very important to detect and investigate the causes of these faults and to apply correct countermeasures.

The Technical Assistance and Support Center (TASC), Nippon Telegraph and Telephone (NTT) East Corporation is engaged in technical consultation and the analysis of optical fiber network faults for the NTT group in Japan and is contributing to eliminating the causes and reducing the number of faults in the optical fiber facilities of FTTH networks. The TASC has investigated and reported faults in various fiber connections using refractive index matching material with wide gaps between fiber ends and faults in fiber connectors with imperfect physical contact [4-6].

This chapter describes some of the faults with optical fiber connections in FTTH networks that the TASC has investigated. In addition, it introduces novel countermeasures for dealing with the faults. The various faults and countermeasures described in this chapter are shown in Fig. 1. First, section 2 briefly reviews a typical FTTH network and various fiber connections in Japan. Then section 3.1 reports faults with fiber connections that employ refractive index matching material. These faults have two major causes: One is a wide gap between fiber ends and the other is incorrectly cleaved fiber ends. Next, section 3.2 describes faults with fiber connections that employ physical contact (PC). This fault has the potential to occur when connector endfaces are contaminated. The characteristics of these faults are outlined. Novel countermeasures against the above-mentioned faults are introduced in section 4. In section 4.1, a new connection method using solid refractive index matching material is proposed as a countermeasure against faults caused by a wide gap between fiber ends. In section 4.2, a fiber optic Fabry-Perot interferometer based sensor is introduced as a way of detecting faults caused by incorrectly cleaved fiber ends. The sensor mainly uses laser diodes, an optical power meter, a 3-dB coupler, and an XY lateral adjustment fiber stage. In section 4.3, a novel tool for inspecting optical fiber ends is proposed as a countermeasure designed to detect faults caused both by incorrectly cleaved fiber ends and contaminated connector endfaces. The proposed tool has a simple structure and does not require focal adjustment. It can be used to inspect a fiber and clearly determine whether it has been cleaved correctly and whether the connector endfaces are contaminated or scratched. This chapter is summarized in section 5.

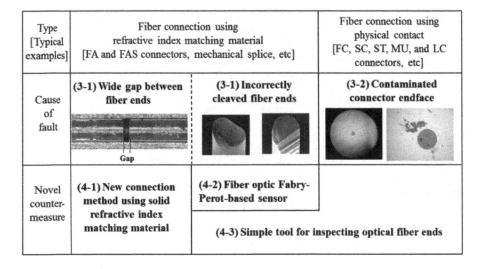

Type [Typical examples]	Fiber connection using refractive index matching material [FA and FAS connectors, mechanical splice, etc]		Fiber connection using physical contact [FC, SC, ST, MU, and LC connectors, etc]
Cause of fault	**(3-1) Wide gap between fiber ends** Gap	**(3-1) Incorrectly cleaved fiber ends**	**(3-2) Contaminated connector endface**
Novel counter-measure	**(4-1) New connection method using solid refractive index matching material**	**(4-2) Fiber optic Fabry-Perot-based sensor**	
		(4-3) Simple tool for inspecting optical fiber ends	

Figure 1. Various faults and their countermeasures dealt with in this chapter

2. Fiber-to-the-home network and various fiber connections

Figure 2 shows the configuration of a typical FTTH network in Japan, which is mainly composed of an optical line terminal (OLT) in a central office, underground and aerial optical fiber cables, and an optical network unit (ONU) inside a customer's home. The network requires various fiber connections at the central office, outdoors, and in homes. With the aerial and home-sited fiber connections in particular, field installable connectors or mechanical splices are used to make it possible to employ the most suitable wiring for the aerial condition and room arrangement. Field assembly (FA) termination connectors and field assembly small (FAS) connectors are types of field installable connectors [7-8].

In contrast, manufactured physical contact (PC)-type connectors, such as miniature-unit coupling optical fiber (MU) and single fiber coupling optical fiber (SC) connectors [9-10], are used in central offices and homes. These connectors require more frequent reconnection than field installable connectors.

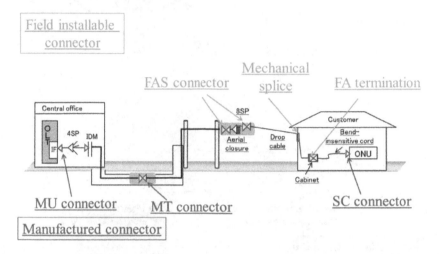

Figure 2. Typical FTTH network and various fiber connections

Figure 3(a) shows the basic structure of a PC-type connector, 3(b) shows that of a mechanical splice and 3(c) shows that of a field installable connector. With PC-type connectors, two ferrules are aligned in an alignment sleeve and connected using compressive force. Normally, two fiber ends in ferrules are connected without a gap and without offset or tilt misalignment. A mechanical splice is suitable for joining optical fibers simply in the field. It consists of a base with a V-groove guide, three coupling plates, and a clamp spring. When a wedge is inserted between the plates and the base, optical fibers can be inserted though the V-groove guide to connect and fix them in position by releasing the wedge between the plates and base [11]. Refractive index matching material is used to reduce Fresnel reflection. This connection procedure requires no electricity.

A field installable connector is composed of three main parts, a polished ferrule containing a short optical fiber (built-in optical fiber), a mechanical splicer, and a clamp. This connector holds the optical fiber drop cable or indoor cable sheath. To assemble the connection, the optical fiber end is cleaved and connected to the built-in optical fiber using a mechanical splice, and the cable sheath is fixed in the clamp. The structure allows connection to another optical fiber connector in the field. In addition, the field installable connector is fabricated based on the above-mentioned mechanical splice technique; therefore, the connection can be assembled without the use of special tools or electricity.

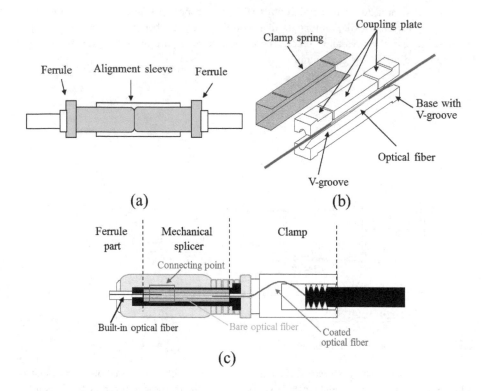

Figure 3. Basic structures of physical contact type connector, (b) mechanical splice, and (c) field installable connector

Both the mechanical splice and the field installable connector use the same fiber end preparation process before fiber installation. Figure 4 shows the fiber end preparation procedures. The coating of a fiber is stripped. Then the stripped fiber (bare fiber) is cleaned with alcohol, cut with a cleaver, inserted into the mechanical splice or the splicer inside a field installable connector, and joined to the opposite fiber or built-in fiber. Finally, the inserted fibers are fixed in position with a clamp. Stripping, cleaning, and cutting are important for successful fiber connection (to provide good performance) in the field. If any of these procedures are not conducted correctly, the performance of the fiber connection might deteriorate.

3. Faults with optical fiber connections

This section reports some of the faults with optical fiber connections in FTTH networks that the TASC has investigated. First, faults related to fiber connection using refractive index matching material are reported in section 3.1. Faults involving PC fiber connection are described in section 3.2.

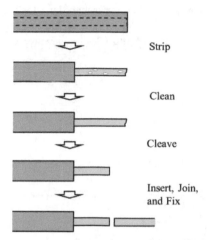

Figure 4. Optical fiber end preparation procedure

3.1. Fiber connection with refractive index matching material

There are two major causes of faults related to fiber connection using refractive index matching material: one is a wide gap between fiber ends and the other is an incorrectly cleaved fiber end. Figure 5 shows three connection models using refractive index matching material; (a) shows the normal connection state with a narrow gap between flat fiber ends, (b) shows an abnormal connection state with a wide gap between flat fiber ends, and (c) shows an abnormal state with an incorrectly cleaved (uneven) fiber end. With the normal connection (a), there is a very narrow (sub-micron) gap between the fiber ends because a normal fiber end is not completely flat. The very narrow gap is filled with silicone oil compound, which is used as the refractive index matching material in a normal connection. In the abnormal connection state (b) there is a very wide gap between the flat fiber ends, and the gap is not filled with refractive index matching material but is a mixed state consisting of refractive index matching material and air. In the abnormal connection state (c) there is a wide gap between flat and incorrectly cleaved (uneven) fiber ends. However, the gap between the fiber ends is filled with matching material.

The optical performance of various fiber connections using refractive index matching material was investigated experimentally. Wide gaps were formed between flat fiber ends by using MT connectors [12] and feeler gauges. A feeler gauge (thickness gauge tape) was installed and fixed in place between the two MT ferrules of a connector with a certain gap size by using a clamp spring. By changing the thickness of the feeler gauge, gaps of various sizes were obtained [13]. In contrast, incorrectly cleaved fiber ends were intentionally formed by adjusting the fiber cleaver so that the bend radius would be too small [14]. The cracks in these incorrectly cleaved fiber ends were from 30 to 200 μm in the axial direction. Using these incorrectly cleaved fiber ends, we fabricated field installable connectors as experimental samples. The fabricated

MT connector with a feeler gauge and field installable connector samples were subjected to a heat-cycle test in accordance with IEC 61300-2-22 (-40 to 70°C, 10 cycles, 6 h/cycle) to simulate conditions in the field. The insertion and return losses were measured.

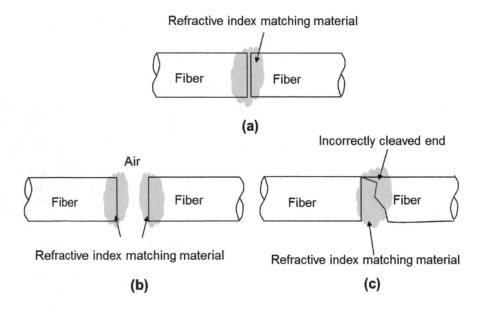

Figure 5. Fiber connection models using refractive index matching material: (a) normal connection with narrow gap between flat fiber ends, (b) abnormal connection with wide gap between flat fiber ends, and (c) abnormal connection with an incorrectly cleaved (uneven) fiber end

The insertion and return losses of an abnormal connection sample with a wide gap between flat fiber ends are shown in Fig. 6. The optical performance changed and was unstable. The insertion loss was initially 2.7 dB and then varied when the temperature changed. The maximum insertion loss exceeded 30 dB. The return losses also varied from 20 dB to more than 60 dB. This performance deterioration is thought to be caused by the mixture of refractive index matching material and air-filled gaps between the fiber ends in the MT connector sample.

Refractive index matching material moved in the gap when the temperature changed, and the mixed state change of the refractive index matching material and the air between the fiber ends induced the change in optical performance. When there is a mixed state consisting of refractive index matching material and air between the fiber ends, the boundary between the refractive index matching material and air could be uneven. In this state, the transmitted light spread randomly in every direction at the boundary. Therefore, the insertion loss increased to more than 30 dB. Consequently, the optical performance of fiber connections with a wide gap between flat fiber ends might be extremely unstable and vary widely. Therefore, it is important to prevent the gap from becoming wider and avoid mixing air with the refractive index matching material in the gap between fiber ends for these fiber connections.

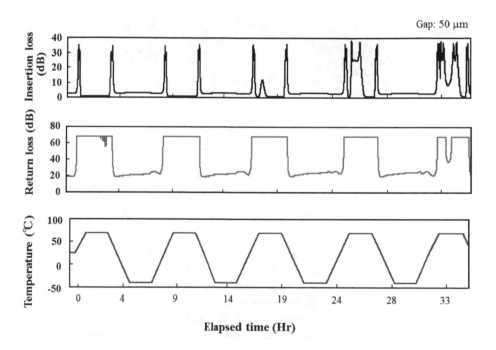

Figure 6. Heat-cycle test results for fiber connection with wide gap between flat fiber ends

The insertion and return losses of an abnormal connection sample with an incorrectly cleaved (uneven) fiber end also changed greatly and were unstable. Figure 7 shows a scatter diagram plotted from the measured insertion and return loss values to enable the values to be easily and simultaneously understood. The horizontal lines indicate insertion loss and the vertical lines indicate return loss. The scatter diagram plots minute insertion and return losses that occurred during the heat cycle test. There are both huge vertical and horizontal fluctuations in the plotted data in Fig. 7. The insertion and return loss values changed periodically during temperature cycles. The initial insertion loss was low at about 1 dB and the initial return loss was high at more than 40 dB. The insertion loss increased greatly and then the return loss decreased as the temperature changed. At worst, the insertion loss changed to 43 dB and the return loss changed to 28 dB.

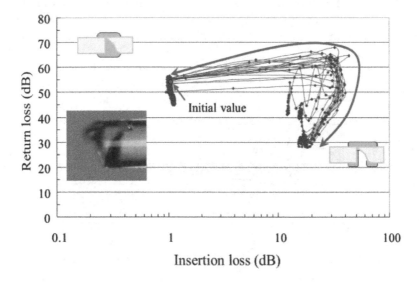

Figure 7. Scatter diagrams of results from heat cycle test for fiber connection with an incorrectly cleaved (uneven) fiber end

The great changes in the insertion and return losses are also attributed to a partially air-filled gap. The gap was not completely filled with refractive index matching material and thus consisted of a mixed state of refractive index matching material and air because of the incorrectly cleaved fiber ends. The boundary between the refractive index matching material and air could be uneven. The transmitted light in this state spread randomly in every direction at the boundary. Therefore, the insertion and return losses became much worse. When the gap was filled with refractive index matching material and there was no air, the optical performance was not so bad. When the gap was a mixed state of refractive index matching material and air, the optical performance deteriorated. The connection state is thought to vary with temperature. These results suggest that the insertion and return losses of fiber connections using incorrectly cleaved fiber ends might change to, at worst, more than 40 dB for the former and less than 30 dB for the latter. Consequently, it is important to prevent gaps between the correctly and incorrectly cleaved ends of fiber connections from becoming wider, and air from mixing with the refractive index matching material in the gaps. Therefore, incorrectly cleaved fiber ends must not be used. An effective countermeasure is to check the fiber ends cleaved with fiber cleavers. Reference [6] is recommended to those readers requiring a more detailed analysis of these abnormal connection states.

3.2. Physical Contact (PC) type connector

This section discusses the deterioration in optical performance caused by the contamination of manufactured physical contact (PC)-type connectors. It has been reported that contamination on a PC-type connector may significantly degrade the performance of mated connectors [15-17]. In this report, contamination was found on the connector endface and the sides of the connector ferrule. To study the effect of contamination on the optical performance of mated connectors, various connection conditions for PC-type connectors in abnormal states are discussed. The abnormal connection conditions are shown in Fig. 8. With PC-type connectors, two ferrules are aligned in an alignment sleeve and connected using compressive force. Normally, two fiber ends in ferrules are connected without a gap and without offset or tilt misalignment. However, if contamination is present, the connection state might become abnormal. An abnormal state can be induced by four conditions: (A) light-blocking caused by contamination on the fiber core, (B) an air-filled gap caused by contamination, (C) tilt misalignment caused by contamination, and (D) offset misalignment caused by contamination. Conditions (A) to (C) are caused by contamination on the ferrule endface. Conditions (C) and (D) are caused by contamination on the side of the ferrule. The performance deterioration caused by contamination (abnormal state) is calculated using the ratio of core contamination coverage and the Marcuse equations [18]. Figure 9 shows the individual calculated insertion losses for the four abnormal conditions. In condition (A), as the core contamination coverage ratio increases, the insertion loss increases. When the ratios are 0.5 and 0.8, the insertion losses are 3 and 7 dB, respectively. This connection condition could degrade the return loss due to the difference between the refractive indices of the fiber core and contamination. Condition (B) may be caused by contamination on the ferrule endface or fiber cladding. As the gap width becomes larger, the calculated insertion loss increases. The insertion loss caused by an air-filled gap is dependent on wavelength. When the wavelengths are 1.31 and 1.55 μm, the insertion losses of a 50-μm gap are 1.0 and 0.4 dB, respectively. This connection condition could also degrade the return loss caused by the difference in the refractive index between the fiber core and air [19]. Condition (C) may be caused by contamination on the edge of the ferrule endface and on the side of the ferrule. As the tilt angle increases, the calculated insertion loss increases. The insertion loss caused by tilt misalignment is dependent on wavelength. When the wavelengths are 1.31 and 1.55 μm, the insertion losses for a 3° misalignment angle are 1.4 and 1.3 dB, respectively. This connection condition might also have a detrimental effect on the return loss due to the difference between the refractive indices of the fiber core and air. Condition (D) may be caused by contamination on the side of the ferrule. When the offset is larger, the calculated insertion loss is higher. The insertion loss caused by offset misalignment is also dependent on wavelength. When the wavelengths are 1.31 and 1.55 μm, the insertion losses of a 3-μm offset are 1.9 and 1.5 dB, respectively. Current PC-type connectors usually have a small clearance between the outer diameter of the ferrule and inner diameter of the alignment sleeve. Therefore, the offset and tilt angle cannot be so large that the insertion losses become low. Conditions (A) and (B), which are caused by contamination on the fiber and ferrule endfaces, are thought to mainly affect the optical performance of connectors.

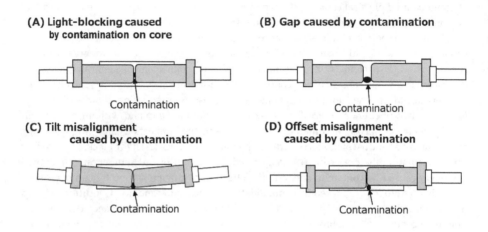

(A) Light-blocking caused
by contamination on core

Contamination

(C) Tilt misalignment
caused by contamination

Contamination

(B) Gap caused by contamination

Contamination

(D) Offset misalignment
caused by contamination

Contamination

Figure 8. Abnormal connection states for PC type connector with contamination

Faults with PC-type connectors caused by contamination were investigated experimentally. Figure 10 shows examples of the investigated connectors. Figure 10 (a) is a normal sample (no contamination on the connector ferrule endface), and (b) to (e) are samples with contamination on the connector ferrule endface. The insertion losses at 1.31 and 1.55 μm were both 0.1 dB and the return loss at 1.55 μm was 58 dB for the uncontaminated sample. This optical performance is good and satisfies the required specifications for an SC connector. However, with the contaminated ferrule endfaces of samples (b) to (d), the insertion losses varied and exceeded 0.5 dB. The return losses were less than 40 dB. This optical performance does not satisfy the specifications for an SC connector. The losses with samples (b) and (c) are thought to be due to condition (A), and the loss with sample (d) is thought to be due to condition (B). With contamination sample (e), the optical performance was not bad and satisfied the SC connector specifications. Consequently, if there is contamination on a PC-type connector, the performance might deteriorate. An effective countermeasure against the loss increase caused by contamination is to inspect the PC-type connector endface prior to connection. When the connector endface is contaminated it must be cleaned with a special cleaner [20]. The countermeasures against connector endface contamination and incorrect cleaving are effective in reducing connector faults.

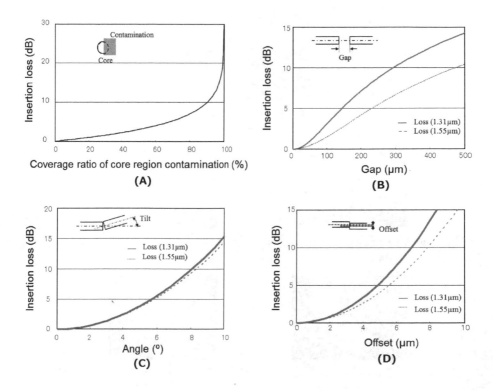

Figure 9. Calculated insertion loss, (A) cover ratio of contamination to fiber core, (B) caused by air-filled gap, (C) caused by tilt, and (D) caused by offset

4. Novel countermeasures

This section introduces novel countermeasures designed to deal with the faults described above. In section 4.1, a new connection method using solid refractive index matching material is proposed as a way of dealing faults caused by wide gaps between fiber ends. In section 4.2, a fiber optic Fabry-Perot interferometer based sensor is introduced as a countermeasure designed to prevent faults caused by incorrectly cleaved fiber ends. In section 4.3, a novel tool for inspecting optical fiber ends is proposed as a technique for detecting both faults caused by incorrectly cleaved fiber ends and those caused by contaminated connector endfaces.

(a) IL=0.1 dB/0.1 dB (1.31 μm/1.55 μm)
RL=58 dB (1.55 μm)

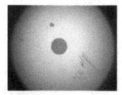   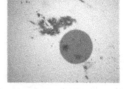

(b) IL=8.7 dB/7.8 dB (1.31 μm/1.55 μm) **(c)** IL=3.3 dB/1.8 dB (1.31 μm/1.55 μm)
RL=34 dB (1.55 μm)   RL=27 dB (1.55 μm)

   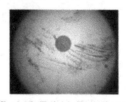

(d) IL=1.05 dB/1.00 dB (1.31 μm/1.55 μm) **(e)** IL=0.27 dB/0.16 dB (1.31 μm/1.55 μm)
RL=25.3 dB (1.55 μm)   RL=51 dB (1.55 μm)

Figure 10. Examples of contamination on connector endface, (a) uncontaminated connector endface, and (b-e) different contaminations on connector endface

4.1. New connection method using solid refractive index matching material

The optical performance of fiber connections with a wide gap between flat fiber ends might be extremely unstable and vary widely. This performance deterioration may not occur immediately after installation but intermittently over time. In the event of an unusual fault, it is difficult to find the defective connection, and it takes long time to repair the fault. Therefore, it is important to prevent the gap between fiber ends from becoming wider in joints that employ refractive index matching material. A novel optical fiber connection method that uses a solid resin as refractive index matching material has been proposed [21]. The new connection method provides a high insertion loss that exceeds the loss budget between network devices when there is a wide gap between fiber ends (defective connection) and a suitable low insertion loss when the gap is less than a particular width (normal connection). The experimental optical performance of the proposed connection method is also discussed in this section.

The following two points are important as regards the new refractive index matching material.

i.   An elastic solid resin must be used that has almost the same refractive index as the fiber core.

ii.   Refractive index matching material with a particular width should be inserted between fiber ends (A and B) and tilted at a special tilt angle to the optical axis of the fiber.

The refractive index matching material must maintain its shape; therefore, a solid resin is used since the connection state cannot be easily changed. Figure 11(a) and (b) show the principles of this connection method. The incident light is refracted at the boundary surface of the refractive index matching material when the fiber ends do not touch it (there is a wide gap between the fiber ends, as shown in Fig. 11(a)). In this case, there is a high insertion loss because of the offset misalignment. In contrast, the incident light will travel straight into the refractive index matching material when it is touched by both fiber ends (the gap between the fiber ends is less than a particular width, as shown in Fig. 11(b)). The insertion loss in Fig. 11(b) is much lower than that in Fig. 11(a).

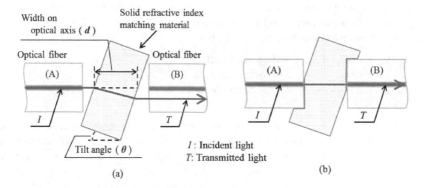

Figure 11. Proposed connection method: (a) fiber ends do not touch matching material, and (b) fiber ends touch matching material

A connection method using solid refractive index matching material based on the above-mentioned considerations was designed and used in the following procedure.

First, a target low insertion loss was set when the gap was narrower than a particular width d and then the particular width of the solid matching material on the optical axis of the fiber was determined.

Then the target high insertion loss was set when the gap was a wider than a particular width d and a special tilt angle θ was determined for the solid refractive index matching material.

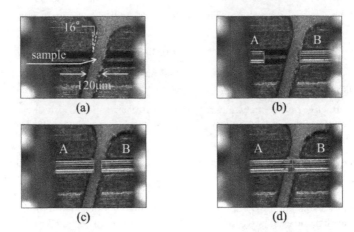

Figure 12. Composition of experimental conditions: (a) V-grooved substrate and sample, (b) fiber A does not touch sample, (c) fiber A just touches sample, and (d) fiber A is close to fiber B and very narrow gap is filled with sample

In step 1, the insertion loss caused by the gap between the fiber ends was calculated by using a Marcuse equation [18]. The insertion loss should be less than 0.5 dB to satisfy the mechanical splice specifications. However, when the insertion loss was 0.5 dB, d was 60 μm, which was too small to handle the refractive index matching material. Therefore, the target d was doubled to 120 μm. The insertion loss then became 2 dB.

In step 2, the insertion loss caused by the misalignment of the offset was calculated by using another Marcuse equation [3]. Another target insertion loss of 20 dB was determined in order to exceed the loss budget between network devices. The insertion loss became 20 dB when θ was 16°.

A sample made of the solid refractive index matching material (silicone resin) was fabricated based on the above parameters. Experiments were carried out with mechanical splices and samples of solid matching material. A groove was dug with the same shape as the sample, and the sample was tilted at 16° to the optical axis of the fiber, as shown in Fig. 12(a). A state was maintained whereby fiber end B always touched the sample, and fiber end A gradually moved toward the sample (Fig. 12(b)-(d)). The insertion and return losses were measured for different gap widths. Figure 12(b) shows the state in which fiber end A did not touch the sample. Figure 12(c) shows the state where fiber end A just touched the sample, and fiber end A was close to fiber end B, and Fig. 12(d) shows the state where the very narrow gap between the fiber ends was filled by the sample.

Figure 13(a) and (b) show the insertion and return loss results at wavelengths of 1.31 and 1.55 μm, respectively. When fiber end A did not touch the sample, the insertion loss always exceeded 20 dB. Moreover, the return losses were constant at 15 dB. When fiber end A just touched the sample, the insertion losses decreased to 2.5 and 2.3 dB, and the return losses increased to 51.7 and 48.6 dB at wavelengths of 1.31 and 1.55 μm, respectively. In addition, when fiber end A was close to fiber end B and the very narrow gap between fiber ends was filled by the sample, both insertion losses decreased to around 0.1 dB, and the return losses were 53.4 and 45.5 dB at wavelengths of 1.31 and 1.55 μm, respectively. These experimental results were consistent with the target values based on the design. If there is a defective connection that has a wide gap, the insertion loss can always be extremely high. In this case, communication services may be immobilized. With the connection method, engineers can detect the defective connection immediately after installation.

Consequently, a new connection method using solid refractive index matching material is proposed as a countermeasure against faults caused by a wide gap between fiber ends. This connection method can provide insertion losses of more than 20 dB or less than 2 dB, respectively, when the gap between the fiber ends is more or less than 120 μm.

4.2. Fiber optic fabry-perot interferometer based sensor

Field installable connections that have incorrectly cleaved fiber ends might lead to insertion losses of more than 40 dB induced by temperature changes, which may eventually result in faults in the optical networks. Therefore, it is important to use correctly cleaved fiber ends to prevent network failures caused by improper optical fiber connections. This means that we need a technique for inspecting cleaved optical fiber ends.

Cleaved optical fiber ends are usually inspected before fusion splicing with a CCD camera and a video monitor installed in fusion splice machines [22]. On the other hand, cleaved optical fiber ends are not usually inspected when mechanical splices and field installable connectors are assembled. These connections are easy to assemble and does not require electric power. Therefore, an inspection method is needed for these connections. A fiber optic Fabry-Perot interferometer based sensor for inspecting cleaved optical fiber ends has been proposed [23-24].

The basic concept of the proposed sensor for inspecting cleaved optical fiber ends is shown in Fig. 14. Figure 14(a) and (b), respectively, show fiber connections in which a fiber with a flattened end for detection is used in the inspection of incorrectly cleaved (uneven) and correctly cleaved (flat) fiber ends. The ratio of the reflected light power (P_r or P_r') to the incident light power (P_i or P_i') within each connection is measured. Two optical fibers are connected with an air gap S remaining between them. Misalignments of the offset and tilt between the fibers and the mode field mismatch are not taken into account. Under both conditions, Fresnel reflections occur at the fiber ends because of refractive discontinuity. In Fig. 14(a), the reflected light from the uneven end spreads in every direction, and the back-reflection efficiency ratio, P_r/P_i, is determined using the Fresnel reflection at the fiber end for detection in air. The Fresnel reflection R_0 is defined by the following equation.

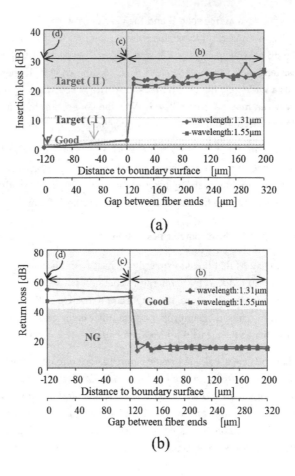

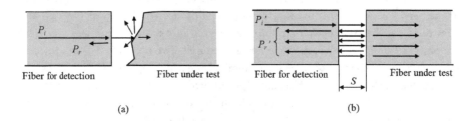

Figure 13. Results of (a) insertion loss and (b) return loss

Figure 14. Basic concept of proposed sensor: (a) inspecting uneven fiber end, and (b) inspecting flat fiber end

$$R_0 = \left(\frac{n_1 - n}{n_1 + n}\right)^2 \tag{1}$$

Here n_1 and n denote the refractive indices of the fiber core and air, respectively.

In Fig. 14(b), some of the incident light is multiply reflected in the gap. The phase of the multiply reflected light changes whenever it is reflected, which interferes with the back-reflected light at the optical fiber connection. These multiple reflections between fiber ends are considered to behave like a Fabry-Perot interferometer [25-27]. Two flat fiber ends make up a Fabry-Perot interferometer. Based on the model, the returned efficiency R ($= P_r'/P_i'$) is defined by the following equation.

$$R = \frac{4R_0 \sin^2(2\pi n S / \lambda)}{(1 - R_0)^2 + 4R_0 \sin^2(2\pi n S / \lambda)} \tag{2}$$

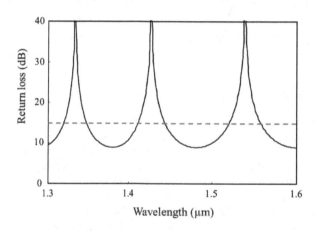

Figure 15. Return losses from uneven (dashed line) and flat (solid line) fiber ends

The return losses in dB are derived by multiplying the log of the reflection functions by -10. Here S and λ denote the gap size and wavelength, respectively. According to Equation (2), the return loss depends on S and λ. Figure 15 shows the calculated return losses from the uneven (incorrectly cleaved) and flat (correctly cleaved) fiber ends. The dashed and solid lines in the figure represent the calculations for the uneven and flat ends based on Equations (1) and (2), respectively. Here, the refractive indices n_1 and n were 1.454 and 1.0, and the gap size used for Equation (2) was 10 μm. The return losses of the uneven end were independent of wavelength and had a constant value of 14.7 dB. The return losses of the flat end varied greatly and periodically and resulted in a worst value of ~8.7 dB because of the Fabry-Perot interference.

The return loss values at wavelengths of 1.31 and 1.55 μm were 11.2 and 18.9 dB, respectively. Even if the gap size and wavelength period were changed, the return losses varied as greatly as the values at a 10-μm gap [28]. These results indicate that an inspected fiber end can be considered uneven or flat depending on whether or not the measured return losses from the fiber end at two wavelengths are both ~14.7 dB.

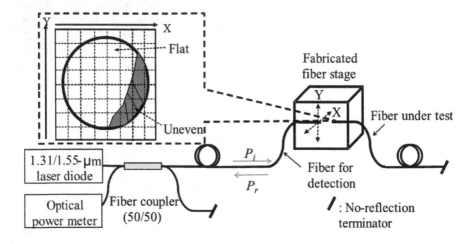

Figure 16. Experimental setup including fiber stage

Based on the above principle, we have designed the inspection sensor shown in Fig. 16. This sensor is composed of two light sources emitting at different wavelengths, an optical power meter, an optical coupler, and a fiber stage. In this proposed sensor, one light source is turned on and the other is turned off. The return loss values are measured separately at two wavelengths. The fiber stage is the most important component because a Fabry-Perot interferometer must be created in it by the fiber for detection and the fiber under test. The other equipment can be adapted from commercially available devices. Therefore, we fabricated a new fiber stage with the following characteristics to implement the proposed technique, as shown in Fig. 17. The dimensions of the fabricated fiber stage are 90 x 100 x 110 mm, which is small enough to be portable in the field. It is also suitable for operation in an outside environment because it does not require a power source. Manual driving was adopted for moving the fiber ends. Two V-grooves for the alignment of two fiber ends were used to create a Fabry–Perot interferometer. These two V-grooves were originally one V-groove that was divided into two. By using the same V-groove for alignment, any tilting of the two fibers along their Z-axes can be reduced. The X- and Y-axes for the scanning direction of the fiber ends were chosen from several alternatives, the direction of the radius, spirally, or with one stroke, due to the streamlining of the fiber stage mechanism. The minimum distance the V-groove can move was designed to be 10 μm along both the X- and Y-axes. The stage was designed to move along both the X- and Y-axes to a maximum distance of 250 μm to cover the entire end of 125-μm-diameter fibers.

Two levers are provided for manually operating only the left V-groove. The left lever moves the left V-groove along the Y-axis at 10 μm per pitch up to a maximum distance of 250 μm. Similarly, the right lever moves the left V-groove along the X-axis at 10 μm per pitch up to a maximum distance of 250 μm.

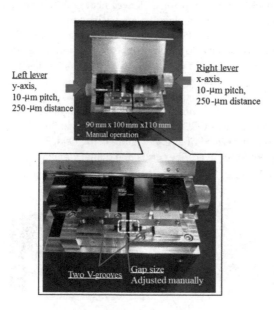

Figure 17. Fabricated fiber stage

In the experiments, the gap between the fiber for detection and the fiber under test was set at 40 μm, and each scanning distance was set at 10 μm. Typical experimental results are shown in Fig. 18. In the figure, (a) and (c) show the flat parts of the inspected fiber end found using the proposed inspection sensor and (b) and (d) show SEM images of the flat end. The fiber ends seen in Fig. 18(a) and (c) were found to be correctly and incorrectly cleaved, respectively. The experimental image with a correctly cleaved fiber end shows that the flat parts form a circle with a diameter of about 140 μm, which is slightly larger than the actual 125-μm-diameter fiber end. This is because the mode field area of light may radiate from the fiber end for detection. In contrast, the experimental results for the incorrectly cleaved fiber end show that half the fiber end parts are flat and half are uneven. The results obtained with the proposed inspection method and those obtained by SEM observation are in good agreement.

The above results show that the proposed sensor made it possible to determine accurately whether the fiber ends were correctly or incorrectly cleaved for all the samples examined. Since the proposed sensor for cleaved optical fiber ends is based on the Fabry-Perot interferometer and a new fiber stage, it allows us to determine whether 10 x 10 μm areas of a cleaved optical fiber end are flat or uneven. The measured results of the inspected flat and uneven fiber ends were in good agreement with those obtained using an SEM.

4.3. Simple tool for inspecting optical fiber ends

The conventional inspection method for a cleaved fiber end involves checking it regularly (about once a week) to ensure good cleaving quality by using a CCD camera and the video monitor of a fusion splicer. If the cleaved fiber end is imperfect, first the fiber cleaver blade is replaced. If no improvements result from this countermeasure, the fiber cleaver itself must be repaired by the manufacturer. In contrast, the conventional inspection method for optical fiber connector endfaces is to check the surface before connecting the mated connector. This method uses a CCD camera and the video monitor of a specialized piece of inspection equipment [29]. If the connector endface is contaminated, it must be cleaned with a special cleaner. These methods using a CCD camera and a video monitor are expensive and unsuitable for use with straightforward fiber connections in the field. Therefore, a simple and economical inspection tool for cleaved fiber ends and connector endfaces suitable for use in the field have been proposed [30-31].

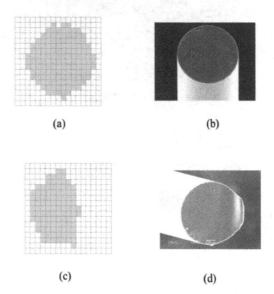

(a)   (b)

(c)   (d)

Figure 18. Experimental results of correctly cleaved fiber end: (a) result with proposed sensor and (b) result of SEM observation, and experimental results for incorrectly cleaved fiber end: (c) result with proposed sensor and (d) result of SEM observation

There are three important requirements for an inspection tool, namely it must provide a clear view, be portable, and easy to operate. We took these requirements into consideration when developing the tool. For the clear view requirement, the fiber ends or connector endfaces under test should be viewable with both the naked eye and a camera. Naked-eye inspection is easily applicable and effective during fiber end preparation and assembly procedures. Camera inspection is effective because it allows us to photograph an inspected cleaved fiber end or connector endface. To meet the portability requirement, the tool must be compact and easy to carry to any location including aerial sites. For the ease of operation requirement, the tool should not require any focal adjustment of a microscope, and the tool must be as easy as possible to handle to prevent the need for complex operations in the field.

Several concrete specifications were determined on the basis of these requirements, as listed in Table 1. The tool must be small enough to carry with one hand. Its total weight should be less than 500 g. It should include a microscope that has a lens with a magnification power of a few hundred times. The target fiber is a 125-μm bare/250-μm coated fiber, which is placed in the FA holder used in field installable connectors or a holder for mechanical splicing. The target connectors are SC, MU, FA, and FAS connectors. The tool uses a cell phone equipped with a CCD camera and small video monitor. This enables the inspected fiber end to be photographed and sent over a cell phone network. LED light sources are used to allow visibility in dark places. A rechargeable battery is used for the LED light sources.

Description	Value/Comment
Size	Small enough to be carried with one hand
Weight	Less than 500 g
Microscope	Few hundred power magnification lens
Target fiber	125/250 (bare/coated) fiber in FA holder or holder for mechanical splicing
Target connector	SC, MU, FA, FAS connectors
Camera	Cell phone capable of taking photos
Light	LED
Battery	Rechargeable battery

Table 1. Specifications of new inspection tool

The tool is designed to inspect both cleaved fiber ends and connector endfaces. Schematic views of the inspection method for a cleaved fiber end and an optical connector endace are shown in Fig. 19(a) to (c). The fundamental optical microscope system for the tool is shown in Fig. 19(a). The microscope system is composed of an objective lens, an eyepiece lens for a cell phone camera or a naked eye, a sample that can be inspected, and an LED light source. Their components must be arranged in a line at designated lengths. In this figure, S_{ob}, L_a, S_{ey}, f_{ob}, and f_{ey} indicate the distance from the objective lens to the object point, the distance between the objective and eyepiece lenses, the distance from the eyepiece lens to the viewpoint for a cell phone camera or the naked eye, the focal distance of the objective lens, and the focal distance of the eyepiece lens, respectively. Here, S_{ob} is designed to be slightly larger than f_{ob} and S_{ey} is designed to be slightly larger than f_{ey}. The figure also shows the light path. An LED light source emitting an almost parallel light beam, is used in this microscope system. After passing through the inspected sample, the light is focused at the back focal plane of the objective lens. It then proceeds to and is magnified by the eyepiece lens before passing into a cell phone camera or a naked eye. The magnified image of the inspected sample can be observed with the cell phone monitor or with the naked eye by using appropriate lenses and by designating appropriate distances; S_{ob}, L_{a}, and S_{ey}. With normal optical microscopes, the inspected sample is placed on the stage and must be adjusted to S_{ob} and aligned at the object point while L_{a} and S_{ey} are designated as constants. By contrast, with this microscope system, the inspected sample, which in placed in a special holder, can always be positioned at the object point without active alignment, i.e., without focal length adjustment. This special holder is described in detail in the following section. For the cleaved fiber inspection shown in Fig. 19(b), the side of the cleaved fiber end is designed to be viewed through the objective lens of the microscope system with the use of the LED light source. The distance between the fiber end and the objective lens a is designed to be equal to S_{ob}. The fiber end, LED, and lens are designed to align passively and to set at each designated distance and not require focal adjustment. However, for the optical-fiber connector inspection in Fig. 19(c), the endface of the connector is designed to be viewed through the objective lens by using a half-mirror and another LED light source. The summation of the distance between the connector endface and the half-mirror b and that between the half-mirror and the object lens c is designed to be equal to S_{ob}. The connector end, LED, half-mirror, and lens are also designed to align passively and to set at each designated distance and not require focal adjustment.

On the basis of the described specifications and design, we developed a simple, mobile and cost-effective tool. The outer components of the proposed inspection tool and the internal makeup of the optical microscope system are shown in Fig. 20. It is composed of three main parts: a body that includes a microscope that has objective and eyepiece lenses and LED light sources, a cell phone and its attachment, and special holders for cleaved fiber ends or connector endfaces. The cell phone is equipped with a CCD camera and a small video monitor. This inspection tool is simple and light, and weighs about 500 g including the cell phone. The optical microscope system is also shown in this figure. The eyepiece lens, objective lens, and object point of the cleaved fiber end are aligned in the body of the tool. The two LEDs for the cleaved fiber end and connector endface are also installed in the body. The half-mirror is aligned in the special holder for the connector endface. The inspection procedure is as follows.

i.   The cleaved optical fiber or the optical connector to be inspected is placed in the appropriate special holder.

ii.   The special holder is installed at the center of the body.

iii.   The attachment with the cell phone is installed on top of the body.

The special holders and body are designed to automatically align the inspected cleaved fiber end or connector end at each of the object points after step (ii). The attachment for a cell phone is also designed to automatically align the camera in the cell phone at the viewpoint after step (iii). This structure and procedure result in the inspection tool not requiring focal adjustment.

The fiber ends or connector endfaces under test can be viewed through the top of the body (step ii) using the cell phone monitor (step iii).

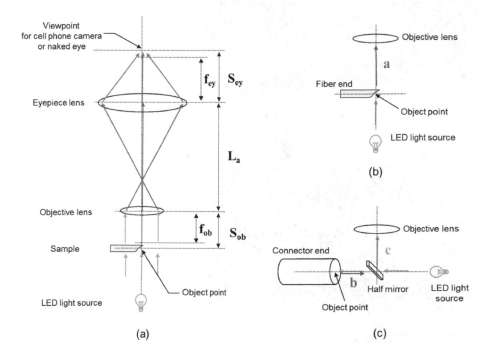

Figure 19. Basic concept of inspection method with developed tool: (a) fundamental optical microscope system and inspecting (b) cleaved fiber end (side) and (c) connector endface (front)

The conventional fiber end preparation procedure for an FAS connector has six steps: (1) cut the support wire of the dropped cable, (2) strip the cable coating, (3) place the fiber in the FA holder, (4) strip the fiber coating, (5) clean the stripped fiber (bare fiber) with alcohol, and (6) cut the bare optical fiber with a fiber cleaver. The assembly procedure comprises the next three steps: (7) insert the properly prepared bare optical fiber into the mechanical splice part in the FAS connector, (8) join it to the built-in optical fiber, and (9) fix the position of the bare optical fiber. The inspection procedure for the proposed inspection tool (i)-(iii) for a cleaved fiber end can be conducted between the fiber end preparation and assembly procedures, i.e., between steps (6) and (7). This indicates that the proposed inspection tool can work well with the conventional fiber end preparation and assembly procedures.

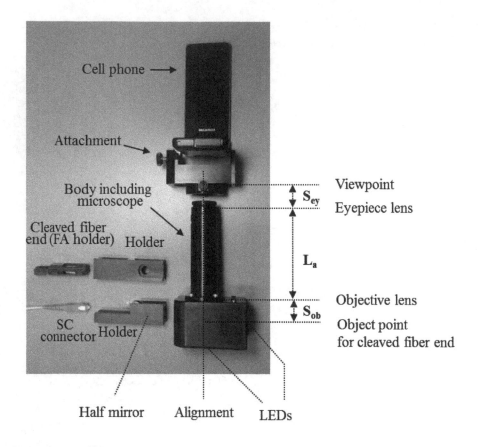

Figure 20. Outer components of fabricated inspection tool and internal makeup of optical microscope system

The inspection results and operation time of the fabricated inspection tool were evaluated. Experimental observation results from the cell phone screen are shown in Fig. 21. The tool with a cell phone attached is shown in Fig. 21(a), and its observation results are shown in Fig. 21(b) to (e). The fiber end or connector endface in each photo is magnified about 100 times. These results indicate that the tool can be used to inspect and determine whether fiber ends have been cleaved incorrectly (Fig. 21(b)) or correctly (Fig. 21(c)), and whether there is contamination (Fig. 21(d)) or no contamination (Fig. 21(e)) on the connector endfaces.

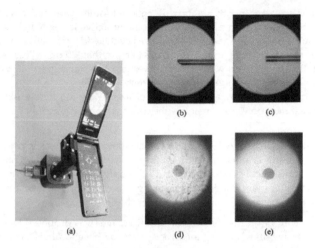

Figure 21. Experimental observation results on cell phone screen: (a) developed inspection tool with cell phone attached, (b) incorrectly cleaved fiber end, (c) correctly cleaved fiber end, (d) contamination on connector endface, and (e) uncontaminated connector endface

For conventional FAS connector procedures, the fiber end preparation and assembly procedures take 72 and 28% of the total installation time, respectively. With the proposed tool, inspection took 11% longer than with the conventional procedure. These results indicate that using the inspection tool may result in a slight increase of 11% in operation time compared with that required with conventional fiber end preparation and assembly procedures.

The fabricated inspection tool is compact, highly portable, and can inspect a fiber and clearly determine whether it has been cleaved correctly and whether contamination or scratches can be found on the connector endfaces. Thus, this tool will be highly practical for field use.

5. Conclusion

This chapter reported example faults and novel countermeasures with optical fiber connectors and mechanical splices in FTTH networks.

After a brief introduction (section 1), section 2 described the FTTH network and optical fiber connectors and mechanical splices used in Japan, and section 3 reported example faults with these optical connections in FTTH networks. First, the faults with fiber connection using refractive-index matching material were reported in section 3.1. There are two major causes of these faults: one is a wide gap between fiber ends and the other is incorrectly cleaved fiber ends. Next, faults with fiber connection using physical contact were explained in section 3.2. This fault might occur when the connector endfaces are contaminated. The characteristics of these faults were outlined.

Novel countermeasures against these above-mentioned faults were introduced in section 4. In section 4.1, a new connection method using solid refractive index matching material was proposed as a countermeasure against faults caused by the wide gap between fiber ends. This connection method can provide an insertion loss of more than 20 dB or less than 2 dB when the gap between the fiber ends is wider or narrower than 120 μm, respectively. If there is a defective connection that has a wide gap, the insertion loss will always be extremely high. In such cases, communication services may be immobilized. With the connection method, engineers undertaking detection work can notice the defective connection immediately after installation.

In section 4.2, a fiber optic Fabry-Perot interferometer-based sensor was introduced as a countermeasure for detecting faults caused by incorrectly cleaved fiber ends. The sensor mainly uses laser diodes, an optical power meter, a 3-dB coupler, and an XY lateral adjustment fiber stage. Experimentally obtained fiber end images were in good agreement with scanning electron microscope observation images of incorrectly cleaved fiber ends.

In section 4.3, a novel tool for inspecting optical fiber ends was proposed as a countermeasure for detecting faults caused both by incorrectly cleaved fiber ends and by contaminated connector endfaces. The proposed tool has a simple structure and does not require focal adjustment. It can be used to inspect a fiber and clearly determine whether it has been cleaved correctly and whether contamination or scratches are present on the connector endfaces. The tool requires a slight increase of 11% in operation time compared with conventional fiber end preparation and assembly procedures. The proposed tool provides a simple and cost-effective way of inspecting cleaved fiber ends and connector endfaces and is suitable for field use.

These results support the practical use of optical fiber connections in the construction and operation of optical network systems such as FTTH.

Acknowledgements

This work was supported by the TASC, NTT East Corporation, Japan. The author is deeply grateful to the following members of the TASC for their support; to Y. Yajima for discussing the experiments related to fiber connection with refractive index matching material, to H. Onose for discussing the experiments related to contaminated connector endfaces, to H. Watanabe and M. Tanaka for helpful discussions regarding the new connection method using solid refractive index matching material and a fiber optic Fabry-Perot interferometer based sensor, and to M. Okada and M. Hosoda for discussions regarding the simple inspection tool.

Author details

Mitsuru Kihara

NTT Access Network Service Systems Laboratories, Nippon Telegraph and Telephone Corporation, Japan

References

[1] Ministry of Internal Affairs and Communications, 2011 WHITE PAPER Information and Communications in Japan [online], Available at: http://www.soumu.go.jp/johotsusintokei/whitepaper/ja/h23/pdf/index.html (accessed 10 Aug. 2012).

[2] Matsuhashi K., Uchiyama Y., Kaiden T., Kihara M., Tanaka H., & Toyonaga M. "Fault cases and countermeasures against damage caused by wildlife to optical fiber cables in FTTH networks in Japan," in Proceeding of the 59th IWCS/Focus, Nov. 7-10, 2010, Providence, Rhode Island, USA, pp. 274-278.

[3] NTT East Corporation, "Fault cases and countermeasures for field assembly connectors in optical access facilities," NTT Technical Review, vol. 9, No. 7, 2011.

[4] Kihara M., Nagano R., Uchino M., Yuki Y., Sonoda H., Onose H., Izumita H., & Kuwaki N., "Analysis on performance deterioration of optical fiber joints with mixture of refractive index matching material and air-filled gaps," in Proceedings of the OFC/NFOEC, March 22-26, 2009, San Diego, CA, USA, JWA4.

[5] Kihara M., Uchino M., Watanabe H., & Toyonaga M., "Fault analysis: optical performance investigation of optical fiber connections with imperfect physical contact," Opt. Lett., vol. 36, no. 24, 2011, pp. 4791-4793.

[6] Kihara M., "Optical performance analysis of single-mode fiber connections," in the book "Optical fiber communications and devices" edited by Moh. Yasin, Sulaiman W. Harun and Hamzah Arof, ISBN 978-953-307-954-7, InTech, February 2, 2012.

[7] Nakajima T., Terakawa K., Toyonaga M., & Kama M., "Development of optical connector to achieve large-scale optical network construction," in Proceedings of the 55th IWCS/Focus, Nov. 12-15, 2006, Providence, Rhode Island, USA, pp. 439–443.

[8] Hogari K., Nagase R., & Takamizawa K., "Optical connector technologies for optical access networks," IEICE Trans. Electron., Vol. E93-C, No. 7, 2010, pp. 1172–1179.

[9] Sugita E., Nagase R., Kanayama K., & Shintaku T., "SC-type single-mode optical fiber connectors," IEEE/OSA J. Lightw. Technol., vol. 7, 1989, pp. 1689–1696.

[10] Nagase R., Sugita E., Iwano S., Kanayama K., & Ando Y., "Miniature optical connector with small zirconia ferrule," IEEE Photon. Technol. Lett., vol. 3, No. 11, 1991, pp. 1045–1047.

[11] "Optical fiber mechanical splice", US patent No.5963699.

[12] Satake T., Nagasawa S., & Arioka R., "A new type of a demountable plastic molded single mode multifiber connector," IEEE/OSA J. Lightw. Technol., vol. LT-4, 1986, pp. 1232-1236.

[13] Kihara M., Nagano R., Izumita H., & Toyonaga M., "Unusual fault detection and loss analysis in optical fiber connections with refractive index matching material," Optical Fiber Technol., vol. 18, 2012, pp. 167–175.

[14] Yajima Y., Watanabe H., Kihara M., & Toyonaga M., "Optical performance of field assembly connectors using incorrectly cleaved fiber ends," in Proceedings of the OECC2011, July 4-8, 2011, Kaohsiung, Taiwan, 7P3_053.

[15] Berdinskikh T., Bragg J., Tse E., Daniel J., Arrowsmith P., Fisenko A., & Mahmoud S., "The contamination of fiber optics connectors and their effect on optical performance," in Technical digest series of OFC 2002, March 17- 22, 2002, Anaheim, California, pp. 617-618.

[16] Albeanu N., Aseere L., Berdinskikh T., Nguyen J., Pradieu Y., Silmser D., Tkalec H., & Tse E., "Optical connector contamination and its influence on optical signal performance," J SMTA, vol. 16, issue 3, 2003, pp. 40-49.

[17] Berdinskikh T., Chen J., Culbert J., Fisher D., Huang S. Y., Roche B. J., Tkalec H., Wilson D., & Ainley S. B., "Accumulation of particles near the core during repetitive fiber connector matings and de-matings" in Technical Digest of OFC/NFOEC2007, March 25-29, 2007, Anaheim, CA, USA, NThA6, pp. 1-11.

[18] Marcuse D., "Loss analysis of optical fiber splice," Bell Sys. Tech. J, vol. 56, 1976, pp. 703-718.

[19] Kihara M., Nagasawa S., & Tanifuji T., "Return loss characteristics of optical fiber connectors," IEEE/OSA J. Lightw. Technol., vol. 14, Sep. 1996, pp. 1986-1991.

[20] For example, CLETOP (Optical connector cleaner), see URL: http://www.ntt-at.com/product/cletop/

[21] Tanaka M., Watanabe H., Kihara M., & Takaya M., "New connection method using solid refractive index matching material," in Proceedings of the OECC2012, July 2-6, 2012, Busan, Korea, 4C2-5.

[22] Tanifuji T., Kato Y., & Seikai S., "Realization of a low-loss splice for single-mode fibers in the field using an automatic arc-fusion splicing machine," in Proceedings of the OFC, Feb. 28-March 2, 1983, New Orleans, Louisiana, USA, MG3.

[23] Kihara M., Watanabe H., Tanaka M., & Toyonaga M., "Fiber optic Fabry-Perot interferometer based sensor for inspecting cleaved fiber ends," Microwave and Optical Technol. Lett., vol. 54, No. 6, 2012, 1541-1543.

[24] Watanabe H., Kihara M., Tanaka M., & Toyonaga M., "Inspection technique for cleaved optical fiber ends based on Fabry-Perot interferometer," IEEE/OSA J. Lightw. Technol., vol. 30, No. 8, 2012, 1244-1249.

[25] Yariv A., "Introduction to optical electronics," New York: Holt, Rinehart, and Winstone, 1985.

[26] Kashima N., "Passive optical components for optical fiber transmission," Norwood, MA: Artech House, 1995.

[27] Kihara M., Tomita S., & Haibara T., "Influence of wavelength and temperature changes on optical performance of fiber connections with small gap," IEEE Photon. Tech. Lett. 18, 2006, 2120-2122.

[28] Kihara M., Uchino M., Omachi M., & Izumita H., "Analyzing return loss deterioration of optical fiber joints with various air-filled gaps over a wide wavelength range," in Proceedings of the OFC/NFOEC, March 21-25, 2010, San Diego, CA, USA, NWE4.

[29] For example, JDSU Fiberscope (Fiber Optic Connector End Face Cleaning System), see URL: http://www.jdsu.com/en-us/Test-and-Measurement/Products/families/westover/Pages/default.aspx/

[30] Okada M., Kihara M., Hosoda M., &Toyonaga M., "Simple inspection tool for cleaved optical fiber ends and optical fiber connector end surface," in Proceedings of the 60th IWCS, Nov. 7-9, Charlotte, North Carolina, USA, 2011, 270-274.

[31] Kihara M., Okada M., Hosoda M., Iwata T., Yajima Y., & Toyonaga M., "Tool for inspecting faults from incorrectly cleaved fiber ends and contaminated optical fiber connector end surfaces," Optical Fiber Technol., vol. 18, 2012, 470-479.

Chapter 4

Multicanonical Monte Carlo Method Applied to the Investigation of Polarization Effects in Optical Fiber Communication Systems

Aurenice M. Oliveira and Ivan T. Lima Jr.

Additional information is available at the end of the chapter

1. Introduction

Polarization-mode dispersion (PMD) is a major source of impairments in optical fiber communication systems. PMD causes distortion and broadens the optical pulses carrying information and lead to inter-symbol interference. In long-haul transmission systems it is necessary to limit the penalty caused by polarization effects [1], so that the probability of exceeding a maximum specified penalty, such as 1 dB, will be small, typically 10^{-5} or less. This probability is referred as the outage probability. Since PMD is a random process, Monte Carlo simulations are often used to compute PMD-induced penalties. However, the rare events of interest to system designers, which consists of large penalties, cannot be efficiently computed using standard (unbiased) Monte Carlo simulations or laboratory experiments. A very large number of samples must be explored using standard unbiased Monte Carlo simulations in order to obtain an accurate estimate of the probability of large penalties, which is computationally costly. To overcome this hurdle, advanced Monte Carlo methods, such as importance sampling (IS) [2], [3] and multicanonical Monte Carlo (MMC) [4] methods, have been applied to compute PMD-induced penalties [5], [6] using a much smaller number of samples. The analytical connections between MMC and IS are presented in [7], [8], [9], [10]. The MMC method has also been used to estimate the bit-error rate (BER) in optical fiber communication systems due to amplified spontaneous emission noise (ASE) [11], for which no practical IS implementation has been developed, and to estimate BER in spectrum-sliced wavelength-division-multiplexed (SS-WDM) systems with semiconductor optical amplifier (SOA) induced noise [12]. More recently, MMC has been used in WDM systems, where the performance is affected by the bit patterns on the various channels and, in order to account for this pattern dependence, a large number of simulations must be performed [13].

In optical fiber communication systems without PMD compensators, the penalty is correlated with the differential group delay (DGD) due to PMD. As a consequence, one can apply IS to bias the DGD [2] for the computation of PMD-induced penalties. However, biasing the DGD alone is inadequate to compute penalties in compensated systems. On the other hand, the use of multiple IS in which both first-and second-order PMD are biased [3] allows one to efficiently study important rare events with large first-and second-order PMD. In [5] and [14], we used multiple IS to bias first-and second-order PMD to compute the outage probability due to PMD in uncompensated systems and in compensated systems with a single-section PMD compensator. The development of IS requires some *a priori* knowledge of how to bias a given parameter in the simulations. In this particular problem, the parameter of interest is the penalty. However, to date there is no IS method that directly biases the penalty. Instead of directly biasing the penalty, one has to rely on the correlation of the first-and second-order PMD with the penalty, which may not hold in all compensated systems. In contrast to IS, MMC does not require *a priori* knowledge of which rare events contribute significantly to the penalty distribution function in the tails, since the bias is done automatically in MMC.

In this chapter, we investigated and applied MMC and IS to accurately and efficiently compute penalties caused by PMD. Using these techniques, we studied the performance of PMD compensators and compared the efficiency of these two advanced Monte Carlo methods to compute the penalty of several types of compensated systems. Since Monte Carlo methods are not deterministic, error estimates are essential to verify the accuracy of the results. MMC is a highly nonlinear iterative method that generates correlated samples, so that standard error estimation techniques cannot be applied. To enable an estimate of the statistical error in the calculations using MMC, we developed a method that we refer to as the MMC transition matrix method [15]. Because the samples are independent in IS simulations, one can successfully apply standard error estimation techniques and first-order error propagation to estimate errors in IS simulations. In this chapter, we also estimate the statistical errors when using MMC and IS. Practical aspects of MMC and IS implementation for optical fiber communication systems are also discussed; in addition, we provide practical guidelines on how MMC can be optimized to accurately and rapidly generate probability distribution functions.

2. MMC Implementation and Estimation of Errors in MMC simulations

In this section, we show how the MMC method can be implemented to PMD emulators and to compute PMD-induced penalty in systems with and without PMD compensators, and also show how one can efficiently estimate errors in MMC simulations using the MMC Transition Matrix method that we developed [15]. For example, when using a standard, unbiased Monte Carlo simulation to calculate the probability density function (pdf) of a statistical quantity, such as the DGD, each sample drawn is independent from the other sample. Hence, when the histogram is smooth, one can infer that the error is acceptably low. The same is not true in MMC simulations because the MMC algorithm requires a substantial de-gree of correlation among the samples to effectively estimate the histogram, which induces a correlation between the calculated values of the probabilities of neighboring bins. Therefore, it is essential to be able to estimate errors particularly in MMC simulation to assess the accuracy of the calculation.

2.1. Multicanonical Monte Carlo method for PMD-Induced penalty

In this sub-section, we briefly review the multicanonical Monte Carlo (MMC) method proposed by Berg and Neuhaus [16], and we describe how we implemented MMC to compute the probability density function (pdf) of the differential group delay (DGD) for PMD emulators. Then, we present results showing the correlation among the histogram bins of the pdf of the DGD that is generated using the MMC method. Finally, we present results with the application of MMC to compute the PMD-induced penalty in uncompensated and single-section compensated system. In particular, we use contours plots to show the regions in the $|\tau|-|\tau_\omega|$ plane that are the dominant source of penalties in uncompensated and single-section PMD compensated systems.

2.1.1. The multicanonical Monte Carlo method

In statistical physics applications, a conventional canonical simulation calculates expectation values at a fixed temperature T and can, by re-weighting techniques, only be extrapolated to a vicinity of this temperature [17]. In contrast, a single multicanonical simulation allows one to obtain expectation values over a range of temperatures, which would require many canonical simulations. Hence, the name multicanonical [16], [17]. The multicanonical Monte Carlo method is an iterative method, which in each iteration produces a biased random walk that automatically searches the state space for the important rare events. Within each iteration, the Metropolis algorithm [18] is used to select samples for the random walk based on an estimated pdf of the quantity of interest or control parameter, which is updated from iteration to iteration. Each new sample in the random walk is obtained after a small random perturbation is applied to the previous sample. In each MMC iteration, a histogram of the control parameter is calculated that records how many samples are in each bin. In each iteration, one generates a pre-determined number of samples that can vary from iteration to iteration. Typically, each iteration has several thousand samples. Once the pre-determined number of samples in any iteration has been generated, the histogram of the control parameter is used to update the estimate of the probability of all the bins as in [16], which will be used to bias the following iteration. After some number of iterations, typically 15 - 50, the number of samples in each bin of the histogram of the control quantity becomes approximately constant over the range of interest, indicating that the estimated pdf of the control quantity is converging to the true pdf.

2.1.2. MMC implementation to PMD emulators

In the computation of the pdf of the DGD, the state space of the system is determined by the random mode coupling between the birefringent sections in an optical fiber with PMD, and the control parameter E is the DGD, as in [19]. When applying MMC, the goal is to obtain an approximately equal number of samples in each bin of the histogram of the control quantity. We compute probabilities by dividing the range of DGD values into discrete bins and constructing a histogram of the values generated by the different random configurations of the fiber sections. The calculations are based on coarse-step PMD emulators consisting of birefringent fiber sections separated by polarization scramblers [20]. We model the fiber using emulators with $N_s = 15$ and $N_s = 80$ birefringent sections. Prior to each section, we use a polarization scrambler to uniformly scatter the polarization dispersion vector on the Poincaré sphere. When polarization scramblers are present, the evolution of the polarization dispersion vector is equivalent to a three-dimensional random walk, and an exact solution [21] is available for the pdf of the DGD that can be compared with the simulations. In unbiased Monte Carlo simulations, the unit matrix $R = R_x(\phi)R_y(\gamma)R_x(\psi)$ rotates the polarization dispersion vector before each section, such that the rotation angles around the x-axis in the i-th section, ϕ_i and ψ_i, have their pdfs uniformly distributed between $-\pi$ and π, while the cosine of the rotation angle γ_i around y-axis has its pdf uniformly distributed between -1 and 1.

Within each MMC iteration, we use the Metropolis algorithm to make a transition from a state k to a state l by making random perturbations $\Delta\phi_i$, $\Delta\gamma_i$, and $\Delta\psi_i$ of the angles ϕ_i, γ_i, and ψ_i in each section, where $\Delta\phi_i$, $\Delta\gamma_i$, and $\Delta\psi_i$ are uniformly distributed in the range $[-\varepsilon\pi, \varepsilon\pi]$. To keep the average acceptance ratio close to 0.5 [22], we choose the coefficient of perturbation $\varepsilon = 0.09$. This perturbation is small, since it does not exceed 10% of the range of the angles. In order to further optimize the MMC simulations and avoid sub-optimal solutions, the random perturbation should also be optimized. We are currently investigating the dependence of the relative error obtained in MMC simulations on the random perturbation and coefficient of perturnation used. The results of this investigation will be published in another publication.

To obtain the correct statistics in γ_i, since in the coarse step method the cosine of γ_i is uniformly distributed, we accept the perturbation $\Delta\gamma_i$ with probability equal to $\min[1, F(\gamma_i + \Delta\gamma_i)/F(\gamma_i)]$, where $F(\gamma) = 0.5(1 - \cos^2\gamma)^{1/2}$. When the perturbation is not accepted, we set $\Delta\gamma_i = 0$. The random variable with acceptance probability given by $\min[1, F(\gamma_i + \Delta\gamma_i)/F(\gamma_i)]$ can be implemented by obtaining a random number from a pdf uniformly distributed between 0 and 1, and then accepting the perturbation $\Delta\gamma_i$ if the random number obtained is smaller than $F(\gamma_i + \Delta\gamma_i)/F(\gamma_i)$. To introduce a bias towards large values of the control parameter E, each transition from state k to the state l in the iteration $j + 1$ is accepted with probability $P_{accept}(k \to l) = \min[1, P^j(E_k)/P^j(E_l)]$, and rejected otherwise, where $P^j(E)$ is the estimate of the pdf of DGD obtained after the first j iterations. At the end of each iteration we update $P^j(E)$ using the same recursion algorithm as in [16], so that the number of hits in each bin of the control parameter histogram becomes approximately equal as the iteration number increases.

2.1.3. Summary of the MMC algorithm

In the first iteration we use M_1 samples and set the pdf of the DGD $P^1(E)$ of a PMD emulator with N_s sections as uniform, $P^1(E)=1/N_b$ (N_b= number of bins). Because every step in the Metropolis algorithm will be accepted with this initial distribution, we more effectively exploit the first iteration by choosing the coefficient of perturbation $\varepsilon=1$ To update the pdf of the DGD at the end of this iteration we use the recursive equation as in (1), which is the same equation used in any other iteration. We then carry out an additional $N-1$ iterations with M_l ($1<l\le N$) samples in each iteration. We note that in general the number of samples in each iteration does not have to be the same. We now present a pseudo-code summary of the algorithm:

Loop over iterations j = 1 to N -1:

  Loop over fiber realizations (samples) m=1 to M_j:

  (1) start random walk on ϕ, γ, and ψ with small steps $\Delta\phi$, $\Delta\gamma$, and $\Delta\psi$

  $\Delta\phi=\{\Delta\phi_1, \cdots, \Delta\phi_{N_s}\}$; $\Delta\gamma=\{\Delta\gamma_1, \cdots, \Delta\gamma_{N_s}\}$; $\Delta\psi=\{\Delta\psi_1, \cdots, \Delta\psi_{N_s}\}$

  (2) compute the provisional value of the DGD ((E_{prov}))

  with the angles $\phi+\Delta\phi$, $\gamma+\Delta\gamma$ and $\psi+\Delta\psi$.

  (3) accept provisional step with probability equal to $\min[1, P^j(E_m)/P^j(E_{prov})]$

  if step accepted: $E_{m+1}=E_{prov}$

  $\phi_{m+1}=\phi_m+\Delta\phi$; $\gamma_{m+1}=\gamma_m+\Delta\gamma$; $\psi_{m+1}=\psi_m+\Delta\psi$

  if step rejected: $E_{m+1}=E_m$

  $\phi_{m+1}=\phi_m$; $\gamma_{m+1}=\gamma_m$; $\psi_{m+1}=\psi_m$

  (4) increment the histogram of E with the sample E_{m+1}

  End of loop over fiber realizations

  update the pdf of the DGD $P^{j+1}(E)$

  restart histogram

  go to next iteration j

End

To update $P^j(E)$ at the end of each iteration j we use the recursive equation [16],

$$P_{k+1}^{j+1} = P_k^{j+1} \frac{P_{k+1}^j}{P_k^j} \left(\frac{H_{k+1}^j}{H_k^j} \right)^{\hat{g}_k^j},$$ (1)

Where $\hat{g}_k^j$ the relative statistical significance of the k-th bin in the j-th iteration, is defined as

Multicanonical Monte Carlo Method Applied to the Investigation of Polarization Effects in
Optical Fiber Communication Systems

77

$$\hat{g}_k^j = \frac{g_k^j}{\sum\limits_{l=1}^{j} g_k^l}, \quad \text{with} \quad g_k^j = \frac{H_{k+1}^j H_k^j}{H_{k+1}^j + H_k^j}. \tag{2}$$

If $H_{k+1}^j + H_k^j = 0$ in a given iteration, then the k-th bin has no statistical significance in this iteration. Therefore, we set $g_k^j = 0$ in that iteration. The statistical significance, $0 \le \hat{g}_k^j \le 1$, depends on both previous bins and previous iterations, inducing a significant correlation among P_k^j. Finally, the P_k^j are normalized so that $\sum\limits_{k=1}^{N_b} P_k^j = 1$, where N_b is the number of bins. MMC is an extension of the Metropolis algorithm [18], where the acceptance rule accepts all the transitions to states with lower probabilities, but rejects part of the more likely transitions to states with higher probabilities. As the number of iterations increases, the histogram of the number of hits in each bin will asymptotically converge to a uniform distribution $\left(H_{k+1}^j / H_k^j \to 1\right)$, and the relative statistical significance will asymptotically converge to zero $\left(\hat{g}_k^j \to 0\right)$. Consequently, P^{j+1} will asymptotically converge to the true probability of the control parameter.

Equations (1) and (2) were derived by Berg and Neuhaus [16] assuming that the probability distribution is exponentially distributed with a slowly varying exponent that is a function of the control quantity (the temperature in their case and DGD or the penalty due to PMD in ours). This assumption is valid in a large number of problems in optical fiber communications, including the pdf of the DGD in fibers with an arbitrary number of sections [19], [23]. The recursions in (1) and (2) were derived by applying a quasi-linear approximation to the logarithm of the pdf in addition to a method for combining the information in the current histogram with that of previous iterations according to their relative statistical significance [16], [19].

2.1.4. Correlations

The goal of any scheme for biasing Monte Carlo simulations, including MMC, is to reduce the variance of the quantities of interest. MMC uses a set of systematic procedures to reduce the variance, which are highly nonlinear as well as iterative and have the effect of inducing a complex web of correlations from sample to sample in each iteration and between iterations. These, in turn, induce bin-to-bin correlations in the histograms of the pdfs. It is easy to see that the use of (1) and (2) generates correlated estimates for the P_k^j, although this procedure significantly reduces the variance [16]. In this section, we illustrate this correlation by showing results obtained when we applied MMC to compute the pdf of the DGD for a PMD emulator with 80 sections.

We computed the correlation coefficient between bin i and each bin j ($1 \le j \le 80$) in the histogram of the normalized DGD by doing a statistical analysis on an ensemble of many independent standard MMC simulations. The normalized DGD, $|\tau| / \langle |\tau| \rangle$, is defined as the

DGD divided by its expected value, which is 30 ps in this case. Suppose that on the l-th MMC simulation, we have P_i^l as the probability of the i-th bin and suppose that the average over all L MMC simulations is $\overline{P}_i$. Then, we define a normalized correlation between bin i and bin j as

$$C(i,j) = \frac{1}{L-1} \sum_{l=1}^{L} \frac{(P_i^l - \overline{P}_i)(P_j^l - \overline{P}_j)}{\sigma_{P_i} \sigma_{P_j}} \tag{3}$$

where σ_{P_i} and σ_{P_j} are the standard deviation of P_i and P_j, respectively. The normalized correlation defined in (3) is known as Pearson's correlation coefficient [24].

The values for $C(i,j)$ generated by (3) will range from -1 to 1. A value of +1 indicates a perfect correlation between the random variables. While a value of -1 indicates a perfect anti-correlation between the random variables. A value of zero indicates no correlation between the random variables.

In Figs. 1–3, we show the correlation coefficients between bin i and bin j, $1 \leq j \leq 80$, for the DGD in the bin i, DGD_i, equal to 30 ps, 45 ps, and 75 ps, respectively. In this case, we used a PMD emulator with 80 sections and the mean DGD is equal to 30 ps. To compute each value of $C(i,j)$ we used $L =32$ MMC simulations. We computed sample mean $\overline{C(i,j)}$ and standard deviation $\sigma_{C(i,j)}$ using 32 samples of $C(i,j)$. The values of the standard deviation for the results shown in Figs. 1–3 are in the range from 1.84×10^{-2} to 3.91×10^{-2}. Note that DGD_i equal to 75 ps represents a case in the tail of the pdf of the DGD, where the unbiased Monte Carlo method has very low probability of generating samples, by contrast to a biased Monte Carlo method such as MMC. The results show that the correlations are not significant until we use a large value for DGD_i compared to the mean DGD. However, these values of DGD_i are precisely the values of greatest interest.

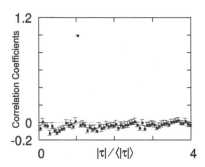

Figure 1. Correlation coefficients between bin i and bin j ($1 \leq j \leq 80$) for the 80-section emulator, where the bin i corresponds to DGD_i=30 ps ($1 \times$ mean DGD). The correlation coefficients are computed using 32 standard MMC simulations. Each standard MMC simulation consists of 30 MMC iterations with 8,000 samples.

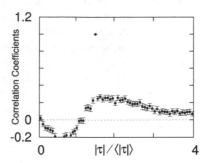

Figure 2. Correlation coefficients between bin i and bin j $(1 \leq j \leq 80)$ for the 80-section emulator, where the bin i corresponds to $DGD_i=45$ ps $(1.5 \times$ mean DGD). The correlation coefficients are computed using 32 standard MMC simulations. Each standard MMC simulation consists of 30 MMC iterations with 8,000 samples.

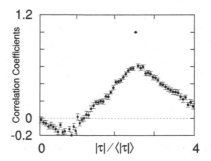

Figure 3. Correlation coefficients between bin i and bin j $(1 \leq j \leq 80)$ for the 80-section emulator, where the bin i corresponds to $DGD_i=75$ ps $(2.5 \times$ mean DGD). The correlation coefficients are computed using 32 standard MMC simulations. Each standard MMC simulation consists of 30 MMC iterations with 8,000 samples.

2.2. Estimation of errors in MMC simulations

In this sub-section, we explain why a new error estimation procedure is needed for multicanonical Monte Carlo simulations, and we then present the transition matrix method that we developed to efficiently estimate the error in MMC. Finally, we present the validation and application of this method.

2.2.1. Why a new error estimation procedure ?

Since MMC is a Monte Carlo technique, it is subject to statistical errors, and it is essential to determine their magnitude. In [25], we showed how to compute errors when using importance sampling. In this sub-section, we show how one can efficiently estimate errors in

MMC simulations using a transition matrix method that we developed. In practice, users of Monte Carlo methods often avoid making detailed error estimates. For example, when using an standard, unbiased Monte Carlo simulation to calculate the pdf of a quantity such as the DGD, the number of samples in each bin of the pdf's histogram is independent. Hence, when the histogram is smooth, one can infer that the error is acceptably low. This procedure is not reliable with MMC simulations because, as we showed in Section 2.1.4, the MMC algorithm induces a high degree of correlation from bin to bin. While it is always best to estimate error with any Monte Carlo method, it is particularly important in MMC simulations, due to the presence of large sample-to-sample correlations on the tails of the distributions.

The existence of correlations in the samples generated with the MMC method makes calculating the errors in MMC simulations significantly more difficult than in standard Monte Carlo simulations. Also, due to the correlations, one cannot apply to MMC standard error analysis that are traditionally used for simulations with uncorrelated samples. For the same reason, one cannot determine the contribution of the variance from each iteration using standard error propagation methods as in the case with importance sampling simulations [5]. Thus, the MMC variance cannot be estimated by applying a standard error analysis to a single MMC simulation. One can in principle run many independent MMC simulations in order to estimate the error by using the standard sample variance formula [26] on the ensemble of MMC simulations. However, estimating the error of the pdf of the quantity of interest by running many independent MMC simulations is computationally costly and in many cases not feasible. One can overcome this problem with the transition matrix method that we developed.

The transition matrix method is an efficient numerical method to estimate statistical errors in the pdfs computed using MMC. In this method, we use the estimated transition probability matrix to rapidly generate an ensemble of hundreds of pseudo-MMC simulations, which allows one to estimate errors from only one standard MMC simulation. The transition probability matrix, which is computed from a single, standard MMC simulation, contains all the probabilities that a transition occurs from any bin of the histogram of the quantity of interest to any other bin after a step (or perturbation) in the MMC random walk. The pseudo-MMC simulations are then made using the computed transition matrix instead of running full simulations. Each pseudo-MMC simulation must be made with the same number of samples per iteration and the same number of iterations as in the original standard MMC simulation. Once an ensemble of pseudo-MMC simulations has been calculated, one can use standard procedures to estimate the error. Since the transition matrix that is used in the pseudo-MMC simulations has its own statistical error, it might seem strange at first that it can be used as the basis from which to estimate the error in the MMC simulations. However, bootstrap theory assures us that such is the case [27]. Intuitively, the variation of any statistical quantity among the members of an ensemble of pseudo-MMC simulations is expected to be the same as the variation among members of an ensemble of standard MMC simulations because the simulations are carried out with the same number of samples and the same number of iterations.

Multicanonical Monte Carlo Method Applied to the Investigation of Polarization Effects in
Optical Fiber Communication Systems

81

To illustrate the transition matrix method, we calculated the pdf of DGD due to PMD and the associated confidence interval for two types of PMD emulators [28]. We validated our method by comparison to the results obtained by using a large ensemble of standard MMC simulations. We tested our method by applying it to PMD emulators because it was the first random phenomenon in optical fiber communication to which MMC was applied [19] and has become essential for testing biasing Monte Carlo methods. Moreover, it is computationally feasible to validate the proposed method with a large ensemble of standard MMC simulations. That is not the case for most other problems, *e.g.*, the error rate due to optical noise [29] and the residual penalty in certain PMD-compensated systems [6].

2.2.2. New error estimation procedure

Here we introduce an efficient numerical procedure that we refer to as the transition matrix method, to compute statistical errors in MMC simulations that properly accounts for the contributions of all MMC iterations. The transition matrix method is a bootstrap resampling method [27], [30] that uses a computed estimate of the probability of a transition from bin i to bin j of the histogram of the DGD. In a bootstrap method, one estimates a complex statistical quantity by extracting samples from an unknown distribution and computing the statistical quantity. In the case of computing the pdf of the DGD in PMD emulators, the complex statistical quantity is the probability of each bin in the histogram of the DGD, the pseudo-samples are the DGD values obtained in the pseudo-MMC simulations, and the unknown distribution is the true transition matrix. One then repeatedly and independently draws an ensemble of pseudo-samples with replacement from each original sample and computes the statistical quantity of interest using the same procedure by which the statistical quantity was first estimated. One can then estimate the variance of the quantity of interest from these pseudo-samples using standard techniques. The bootstrap method is useful when it is computationally far more rapid to resample the original set of samples than to generate new samples, allowing for an efficient estimate of the variance.

2.3. Bootstrap method

Efron's bootstrap [27] is a well-known general purpose technique for obtaining statistical estimates without making *a priori* assumptions about the distribution of the data. A schematic illustration of this method is shown in Fig.4. Suppose one draws a random vector $x=(x_1,x_2...,x_n)$ with n samples from an unknown probability distribution F and one wishes to estimate the error in a parameter of interest $\hat{\theta}=f(x)$. Since there is only one sample of $\hat{\theta}$, one cannot use the sample standard deviation formula to compute the error. However, one can use the random vector x to determine an empirical distribution $\hat{F}$ from F (unknown distribution). Then, one can generate bootstrap samples from $\hat{F}$, $x^*=(x_1^*,x_2^*,...,x_n^*)$, to obtain $\hat{\theta}^*=f(x^*)$ by drawing n samples with replacement from x. The quantity $f(x^*)$ is the result of applying the same function $f(.)$ to x^* as was applied to x. For example, if $f(x)$ is the median of x, then $f(x^*)$ is the median of the bootstrap resampled data set. The star notation indicates that x^* is not the actual data set x, but rather a resampled version of x obtained from the estimated distribution $\hat{F}$. Note that one can rapidly generate as many bootstrap samples $\mathbf{x}^*$ as one needs, since those simulations do not make use the system model, and then generate independent bootstrap sample estimates of $\hat{\theta}$, $\hat{\theta}_1^* = f(\mathbf{x}_1^*)$, ... , $\hat{\theta}_B^* = f(\mathbf{x}_B^*)$, where B is the total number of bootstrap samples. Then, one can estimate the error in $\hat{\theta}$ using the standard deviation formula on the bootstrap samples $\hat{\theta}^*$.

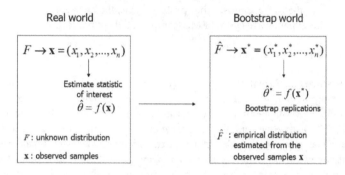

Real world   Bootstrap world

$$F \rightarrow \mathbf{x} = (x_1, x_2, ..., x_n)$$ $$\hat{F} \rightarrow \mathbf{x}^* = (x_1^*, x_2^*, ..., x_n^*)$$

Estimate statistic
of interest
$$\hat{\theta} = f(\mathbf{x})$$ $$\hat{\theta}^* = f(\mathbf{x}^*)$$
                                                    Bootstrap replications

F: unknown distribution $\hat{F}$: empirical distribution
                                                    estimated from the
$\mathbf{x}$: observed samples   observed samples $\mathbf{x}$

Figure 4. On the left, we show the drawing of a true realization form the actual, unknown distribution F. On the right, we show the same procedure applied to drawing bootstrap realizations.

The transition matrix method that we describe in this chapter is related to the bootstrap re-sampling method as follows:

1. $\hat{F}$ is an estimate of the transition matrix obtained from a single standard MMC simulation;

2. $\mathbf{x}_1^*, ..., \mathbf{x}_B^*$, are the collection of samples that is obtained from the ensemble of pseudo-MMC simulations. We note that $\mathbf{x}_b^*$ should be computed using the exact same number of iterations and the exact same number of samples per iteration as in the original standard MMC simulation;

3. Each $\hat{\theta}_b^*$, where $b=1,2,...,B$, is a value for the probability p_k^* of the k-th bin of the histogram of the DGD obtained from each of the pseudo-MMC simulations;

4. Given that one has B independent p_k^*, one can obtain an error estimate for each bin in the estimated pdf of the DGD using the traditional sample standard deviation formula [26, 27].

$$\sigma_{\hat{\theta}^*} = \left[\frac{1}{B-1} \sum_{b=1}^{B} \left(\hat{\theta}_b^* - \overline{\hat{\theta}^*} \right)^2 \right]^{1/2},$$ (4)

where,

Multicanonical Monte Carlo Method Applied to the Investigation of Polarization Effects in
Optical Fiber Communication Systems

83

$$\overline{\theta^*} = \frac{1}{B}\sum_{b=1}^{B}\hat{\theta}_b^*.$$ (5)

2.4. The transition matrix method

In this sub-section, we explain the transition matrix method in the context of computing errors in the pdf of the DGD for PMD emulators. The transition matrix method has two parts. In the first part, one obtains an estimate of the pdf of the DGD and an estimate of the one-step transition probability matrix Π. To do so, one runs a standard MMC simulation, as described in Section 2.1.2. At the same time, one computes an estimate of the transition probability $\pi_{i,j}$, which is the probability that a sample in the bin i will move to the bin j after a single step in the MMC algorithm. We stress that a transition attempt must be recorded whether or not it is accepted by the Metropolis algorithm after the fiber undergoes a random perturbation. The transition matrix is a matrix that contains the probability that a transition will take place from one bin to any other bin when applying a random perturbation. It is independent of the procedure for rejecting or accepting samples, which is how the biasing is implemented in the MMC method. An estimate of the transition matrix that is statistically as accurate as the estimate of the pdf using MMC can be obtained by considering all the transitions that were attempted in the MMC ensemble. One uses this information to build a $N_b \times N_b$ one-step transition probability matrix, where N_b is the number of bins in the histogram of the pdf. The transition matrix Π consists of elements $\pi_{i,j}$, where the sum of the row elements of Π equals 1. The elements $\pi_{i,j}$ are computed as

$$\pi_{i,j} = \frac{\sum_{m=1}^{M_t-1} I_i(E_m)I_j(E_{m+1})}{\sum_{m=1}^{M_t-1} I_i(E_m)}, \text{ if } \sum_{m=1}^{M_t-1} I_i(E_m) \neq 0,$$ (6)

And $\pi_{i,j} = 0$, otherwise. In (6), M_t is the total number of samples in the MMC simulation and E_m is the m-th DGD sample. The indicator function $I_i(E)$ is chosen to compute the probability of having a DGD sample inside the bin i of the histogram. Thus, $I_i(E)$ is defined as 1 inside the DGD range of the bin i, otherwise $I_i(E)$ is defined as 0. In the second part of the procedure, one carries out a new series of MMC simulations (using the transition probability matrix), that we refer to as pseudo-MMC simulations. In each step, if one starts for example in bin i of the histogram, one picks a new provisional bin j using a procedure to sample from the pdf π_i, where $\pi_i(j) = \pi_{i,j}$. One then accepts or rejects this provisional transition using the same criteria as in full, standard MMC simulations, and the number of samples in the bins of histogram is updated accordingly. Thus, one is using the transition matrix Π to emulate the random changes in the DGD that result from the perturbations $\Delta\phi_i$, $\Delta\gamma_i$, and $\Delta\psi_i$ that were used in the original standard MMC simulation. In all other respects, each pseudo-MMC simulation is like the standard MMC simulation. In particular, the metric for

accepting or rejecting a step, the number of samples per iteration, and the number of iterations must be kept the same. It is possible to carry out hundreds of these pseudo-MMC simulations in a small fraction of the computer time that it takes to carry out a single standard
MMC simulation. This procedure requires us to hold the entire transition matrix in memory,
which could in principle be memory-intensive, although this issue did not arise in any of the
problems that we considered. This procedure will be useful when evaluating a transition using the transition matrix requires far less computational time than calculating a transition
using the underlying physics. This is an assumption that was valid for the cases in which we
considered, and we expect that it is applicable to most practical problems. An estimate of the
pdf of the DGD is obtained in the final iteration of each pseudo-MMC simulation. Since the
estimates of the probability in a given bin in the different pseudo-MMC simulations are independent, one may apply the standard formula for computation of the variance $\sigma_{p_i}^2$ of the
i-th bin

$$\sigma_{p_i}^2 = \frac{1}{(B-1)}\sum_{b=1}^{B}\left(p_{i,b}^* - \overline{p_i^*}\right)^2 , \quad \text{with} \quad \overline{p_i^*} = \frac{1}{B}\sum_{b=1}^{B}p_{i,b}^* , \tag{7}$$

where $p_{i,b}^*$ is the probability of the i-th bin in the histogram of the DGD obtained in the b-th
pseudo-MMC simulation and B is the total number of pseudo-MMC simulations. Thus, $\sigma_{p_i^*}$
is an estimate of the error in the i-th bin in the histogram of the DGD obtained in a single
MMC simulation. We now illustrate the details of how we choose the provisional transition
from bin i to bin j with the following pseudo-code:

```
bin DGD of current sample = i
use random number to generate x from a uniform pdf between 0 and 1: x ← U[0,1]
for j=1 to N_b
        if (x < π_{i,j}^{cdf})
                        new bin = j
                        break
        end if
    end for
current bin = new bin
```

where $\pi_{i,j}^{cdf} = \sum_{m=1}^{j}\pi_{i,m}$ is the cumulative transition probability. This procedure is used to sample
from the pdf π_i, where $\pi_i(j) = \pi_{i,j}$, and with $\pi_{i,j}$ defined as the probability that a sample in
the bin i will move to the bin j.

2.5. Assessing the error in the MMC error estimation

The estimate of the MMC variance also has an error, which depends on the number of samples in a single standard MMC simulation and on the number of pseudo-MMC simulations (bootstrap samples) [31]. Here, the error due to the bootstrap resampling is minimized by using 1,000 bootstrap pseudo-MMC simulations. Therefore, the residual error is due to the finite number of samples used to estimate both the pdf of the DGD and the transition matrix in the single standard MMC simulation, *i.e.*, in the first part of the transition matrix method. Thus, there is a variability in the estimate of the MMC variance due to the variability of the transition matrix $\hat{\Pi}$ as an estimate of the true transition matrix Π. To estimate the error in the estimate of the MMC variance, we apply a procedure known in the literature as *bootstrapping the bootstrap* or *iterated bootstrap* [32]. The procedure is based on the principle that if the bootstrap can estimate errors in one statistical parameter using $\hat{\Pi}$, one can also use bootstrap to check the uncertainty in the error estimate using bootstrap resampled transition matrices $\hat{\Pi}^*$. The procedure consists of:

1. Running one standard MMC simulation;

2. Generating N_B=100 pseudo-MMC simulations and computing transition matrices for each of the pseudo-MMC simulation. Therefore, we obtain N_B transition matrices that we call pseudo-transition matrices $\hat{\Pi}^*_B$;

3. For each pseudo-transition matrix $\hat{\Pi}^*_B$ we calculate N_B=100 pseudo-MMC simulations (N_B values for the probability of any given bin of the estimated pdf of the DGD, p^{**}). The double star notation indicates quantities computed with bootstrap resampling from a pseudo-transition matrix. We then estimate the error for the probability of any given bin in the estimated pdf of the DGD, $\sigma_{p^{**}}$, for each pseudo-transition matrix;

4. Since we have N_B=100 pseudo-transition matrices, we repeat step 3 N_B times and obtain N_B values for $\sigma_{p^{**}}$. Then, we compute the double bootstrap confidence interval Δp^{**} of the relative variation of the error of p (statistical error in p, where p is the probability of any given bin in the estimated pdf of the DGD computed using a single standard MMC simulation):

$$\Delta p^{**} = \left[\frac{\overline{\sigma_{p^{**}}} - \sigma_{\left(\sigma_{p^{**}}\right)}}{p}, \frac{\overline{\sigma_{p^{**}}} + \sigma_{\left(\sigma_{p^{**}}\right)}}{p} \right], \tag{8}$$

where,

$$\sigma_{\left(\sigma_{p}^{**}\right)} = \left[\frac{1}{N_B - 1} \sum_{n=1}^{N_B} \left(\sigma_{p}^{(n)**} - \overline{\sigma_{p}^{**}}\right)^2\right]^{1/2}, \tag{9}$$

and

$$\overline{\sigma_{p}^{**}} = \frac{1}{N_B} \sum_{n=1}^{N_B} \sigma_{p}^{(n)**}. \tag{10}$$

In (9) and (10), $\sigma_{p}^{(n)**}$ is the standard deviation of p^{**} computed using the n-th pseudo-transition matrix.

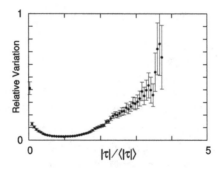

Figure 5. Relative variation $(\hat{\sigma}_{\hat{P}_{DGD}} / \hat{P}_{DGD})$ of the pdf of the normalized DGD, $|\tau| / \langle|\tau|\rangle$, for the 15-section PMD emulator using 14 MMC iterations with 4,000 samples. The confidence interval is given by (8) when we compute an ensemble of standard deviations using bootstrap resampling for each of the 100 pseudo-transition matrices.

In Fig. 5, we show the relative variation of p^{**} and its confidence interval Δp^{**} for a PMD emulator with 15 sections. We used 14 MMC iterations with 4,000 samples each (total of 56,000 samples). The confidence interval of the relative variation is defined in (8). We used a total of 80 evenly-spaced bins where we set the maximum value for the normalized DGD as five times the mean DGD. We also use the same number for bins for all the figures shown in this chapter. As expected, we observed that the error in the estimate of the MMC variance is large when the MMC variance is also large. The confidence interval Δp^{**} is between $(2.73 \times 10^{-2}, 3.19 \times 10^{-2})$ and $(3.61 \times 10^{-1}, 4.62 \times 10^{-1})$ for $|\tau| / \langle|\tau|\rangle < 2$. It increases to $(2.68 \times 10^{-1}, 4.48 \times 10^{-1})$ when $|\tau| / \langle|\tau|\rangle = 3$ and to $(4.05 \times 10^{-1}, 9.09 \times 10^{-1})$ at the largest value of $|\tau| / \langle|\tau|\rangle$. We concluded that the estimate of the relative variation of the probability of a bin is a good estimate of its own accuracy. This result is similar to what is observed with the standard analysis of standard Monte Carlo simulations [26]. Intuitively, one expects the relative error and the error in the estimated error to be closely related because both are drawn from the same sample space. In Fig. 5, we also observe that the relative variation increases with the DGD for values larger than the mean DGD, especially in the tail of the pdf. This phenomenon occurs because the regions in the configuration space that contribute to the tail of the pdf of the DGD are only explored by the MMC algorithm after several iterations. As the number of iterations increases, the MMC algorithm allows the exploration of less probable regions of the configuration space. Because less probable regions are explored in the last iterations, there will be a significantly smaller number of hits in the regions that contribute to the tail of the pdf of the DGD. As a consequence, the relative variation will increase as the DGD increases.

2.6. Application and validation

We estimated the pdf of the normalized DGD $(\hat{P}_{\text{DGD}})$ and its associated confidence interval $\Delta \hat{P}_{\text{DGD}}$ for PMD emulators comprised of 15 and 80 birefringent fiber sections with polarization scramblers at the beginning of each section. The normalized DGD, $|\tau|/\langle|\tau|\rangle$, is defined as the DGD divided by its expected value, which is equal 30 ps. We used 14 MMC iterations with 4,000 samples each to compute the pdf of the normalized DGD when we used a 15-section emulator and 30 MMC iterations with 8,000 samples each when we used an 80-section PMD emulator.

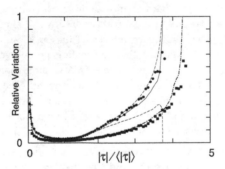

Figure 6. Relative variation $(\hat{\sigma}_{\hat{P}_{\text{DGD}}}/\hat{P}_{\text{DGD}})$ of the pdf of the normalized DGD, $|\tau|/\langle|\tau|\rangle$. (i) Circles: Transition matrix method based on a single standard MMC simulation for the 15-section PMD emulator; (ii) Solid: 10^3 standard MMC simulations for the 15-section emulator; (iii) Dashed: Confidence interval of the relative variation of the error estimated using the transition matrix method for the 15-section PMD emulator; (iv) Squares: Transition matrix method based on a single standard MMC simulation for the 80-section PMD emulator; (v) Dot-dashed: 10^3 standard MMC simulations for the 80-section PMD emulator.

We monitored the accuracy of our computation by calculating the relative variation of the pdf of the normalized DGD. The relative variation is defined as the ratio between the standard deviation of the pdf of the normalized DGD and the pdf of the normalized DGD $(\hat{\sigma}_{\hat{P}_{\text{DGD}}}/\hat{P}_{\text{DGD}})$. In Fig. 6, we show the relative variation when we used PMD emulators with

15 and with 80 birefringent sections. The symbols show the relative variation when we applied the procedure that we described in Section 2 with 1,000 pseudo-MMC simulations based on a single standard MMC simulation and the transition matrix method, while the solid and the dot-dashed lines show the relative variation when we used 1,000 standard MMC simulations. The circles and the solid line show the results for a 15-section PMD emulator, while the squares and dot-dashed line show the results when we used an 80-section PMD emulator. As expected, the result from an ensemble of pseudo-MMC simulations shows a systematic deviation from the result from an ensemble of standard MMC simulations for both emulators. The systematic deviation changes depending on which standard MMC simulation is used to generate the pseudo ensemble. In Fig. 6, the two dashed lines show the confidence interval of the relative variation with the 15-section PMD emulator computed using the transition matrix method, i.e., the confidence interval for the results that are shown with the circles. The confidence interval Δp^{**} is between $(3.04\times10^{-2}, 3.28\times10^{-2})$ and $(2.76\times10^{-1}, 3.62\times10^{-1})$ for $|\tau|/\langle|\tau|\rangle<2$. It increases to $(2.39\times10^{-1}, 4.31\times10^{-1})$ when $|\tau|/\langle|\tau|\rangle=3$ and to $(2.69\times10^{-1}, 9.88\times10^{-1})$ at the largest value of $|\tau|/\langle|\tau|\rangle$.

While the relative variation that is computed using the transition matrix method from a single MMC simulation will vary from one standard MMC simulation to another, the results obtained from different standard MMC simulations are likely to be inside this confidence interval with a well-defined probability. The confidence interval of the relative variation was obtained using a procedure similar to the one discussed in the Section 2.2, except that we computed the relative variation of the probability of a bin using the transition matrix method for every one of the 1,000 standard MMC simulations. Therefore, we effectively computed the true confidence interval of the error estimated using the transition matrix method. We have verified that the confidence interval calculated using the double bootstrap procedure on a single standard MMC simulation agrees well with the true confidence interval in all the cases that we investigated. We observed an excellent agreement between the results obtained with the transition matrix method based on a single standard MMC simulation and the results obtained with 1,000 standard MMC simulations for both 15 and 80 fiber sections when the relative variation $(\hat{\sigma}_{\hat{P}_{\text{DGD}}}/\hat{P}_{\text{DGD}})$ is smaller than 15%. For larger relative variation, the true error is within the confidence interval of the error, which can be estimated using the double bootstrap method described in Section 2.2. The curves for the 80-section PMD emulator have a larger DGD range because a fiber with 80 birefringent sections is able to produce larger DGD values than is possible with a fiber with 15 birefringent fiber sections [28].

In Figs. 7 and 8, we show with symbols the results for the pdf of the normalized DGD and its confidence interval using the numerical procedure that we presented in Section 2.2. The solid line shows the pdf of the normalized DGD obtained analytically using a solution (see [21]) for 15 and 80 concatenated birefringent fiber sections with equal length. For comparison, we also show the Maxwellian pdf for the same mean DGD. In table 1, we present selected data points from the curves shown in Fig. 7. For both 15- and 80-section emulators, we find that the MMC yields estimates of the pdf of the normalized DGD with a small confi-

Multicanonical Monte Carlo Method Applied to the Investigation of Polarization Effects in Optical Fiber Communication Systems

89

dence interval. In Figs. 7 and 8, we see that the standard deviation $\left(\hat{\sigma}_{\hat{P}_{DGD}}\right)$ for the DGD pdf is always small compared to the DGD pdf. The values of the relative variation $\left(\hat{\sigma}_{\hat{P}_{DGD}} / \hat{P}_{DGD}\right)$ ranges from 0.016 to 0.541. We used only 56,000 MMC samples to compute the pdf of the DGD in a 15-section emulator, but we were able nonetheless to accurately estimate probabilities as small as 10^{-8}. Since the relative error in unbiased Monte Carlo simulations is approximately given by $N_I^{-1/2}$, where N_I is the number of hits in a given bin, it would be necessary to use on the order of 10^9 unbiased Monte Carlo samples to obtain a statistical accuracy comparable to the results that I show in the bin with lowest probability in Figs. 7 and 8.

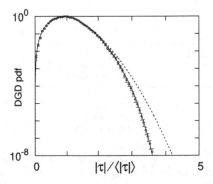

Figure 7. The pdf of the normalized DGD, $|\tau| / \langle|\tau|\rangle$, for the 15-section PMD emulator using 14 MMC iterations with 4,000 samples. (i) Diamonds: DGD pdf with error estimation using the transition matrix method, (ii) Dashed line: Maxwellian pdf, (iii) Solid line: Analytical pdf of the DGD for the 15-section PMD emulator.

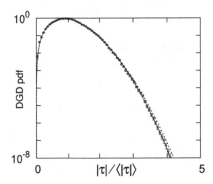

Figure 8. The pdf of the normalized DGD, $|\tau| / \langle|\tau|\rangle$, for the 80-section PMD emulator using 30 MMC iterations with 8,000 samples. (i) Diamonds: DGD pdf with error estimation using the transition matrix method, (ii) Dashed line: Maxwellian pdf, (iii) Solid line: Analytical pdf of the DGD for the 80-section PMD emulator.

$\lvert\tau\rvert/\langle\lvert\tau\rvert\rangle$	P_{DGD}	$\hat{P}_{DGD}$	$\hat{\sigma}_{\hat{P}_{DGD}}$	$\hat{\sigma}_{\hat{P}_{DGD}}/\hat{P}_{DGD}$
0.031	3.00×10^{-3}	4.50×10^{-3}	1.35×10^{-3}	0.301
0.344	3.16×10^{-1}	2.84×10^{-1}	1.75×10^{-2}	0.062
0.719	8.56×10^{-1}	8.57×10^{-1}	2.83×10^{-2}	0.033
1.094	8.63×10^{-1}	8.50×10^{-1}	2.76×10^{-2}	0.033
1.469	4.64×10^{-1}	4.66×10^{-1}	2.16×10^{-2}	0.046
1.844	1.43×10^{-1}	1.36×10^{-1}	1.21×10^{-2}	0.089
2.219	2.50×10^{-2}	2.32×10^{-2}	3.37×10^{-3}	0.145
2.594	2.26×10^{-3}	2.15×10^{-3}	4.43×10^{-4}	0.206
2.969	8.70×10^{-5}	7.57×10^{-5}	2.16×10^{-5}	0.286
3.344	8.92×10^{-7}	8.13×10^{-7}	3.49×10^{-7}	0.430
3.594	1.10×10^{-8}	1.59×10^{-8}	8.63×10^{-9}	0.541

Table 1. Selected data points from the curves shown in Fig. 6. The columns from left to right show: the normalized DGD value, the analytical probability density function, the estimated probability density function, the standard deviation computed using the transition matrix method, and the relative variation.

We would like to stress that the computational time that is required to estimate the errors using the transition matrix method does not scale with the time needed to carry out a single standard MMC simulation. For instance, it takes approximately 17.5 seconds of computation using a Pentium 4.0 computer with 3 GHz of clock speed to estimate the errors in the pdf of the DGD for the 80-section emulator using 1,000 pseudo-MMC simulations with the transition matrix method, once the transition matrix is available. The computational time that is required to compute the pdf of the DGD using only one standard MMC simulation is 60 seconds. To obtain 1,000 standard MMC simulations would require about 16.6 hours of CPU time in this case.

We also stress that it is difficult to estimate the statistical errors in MMC simulations because the algorithm is iterative and highly nonlinear. We introduced the transition matrix method that allows us to efficiently estimate the statistical errors from a single standard MMC simulation, and we showed that this method is a variant of the bootstrap procedure. We applied this method to calculate the pdf of the DGD and its expected error for 15-section and 80-section PMD emulators. Finally, we validated this method in both cases by comparing the results to estimates of the error from ensembles of 1,000 independent standard MMC simulations. The agreement was excellent. In Section 4, we apply the transition matrix method to estimate errors in the outage probability of PMD uncompensated and compensated systems. We anticipate that the transition matrix method will allow one to estimate errors with any application of MMC including the computation of the pdf of the received voltage in optical communication systems [29] and the computation of rare events in coded communication systems [33].

3. PMD Compensators

In this chapter, we investigated a single-section and three-section PMD compensators. A single-section PMD compensator [34], which is a variable-DGD compensator that was programmed to eliminate the residual DGD at the central frequency of the channel after compensation, and a three-section PMD compensator proposed in [35], which compensates for first- and second-order PMD. The three-section compensator consists of two fixed-DGD elements that compensate for the second-order PMD and one variable-DGD element that eliminates the residual DGD at the central frequency of the channel after compensation. The three-section compensator that we used has the first- and second-order PMD as feedback parameters. This compensator can also in principle operate in a feedforward configuration.

3.1. Single-section compensator

The increased understanding of PMD and its system impairments, together with a quest for higher transmission bandwidths, has motivated considerable effort to mitigate the effects of PMD, based on different compensation schemes [36], [37], [38]. One of the primary objectives has been to enable system upgrades from 2.5 Gbit/s to 10 Gbit/s or from 10 Gbit/s to 40 Gbit/s on old, embedded, high-PMD fibers. PMD compensation techniques must reduce the impact of first-order PMD and should reduce higher-order PMD effects or at least not increase the higher orders of PMD. The techniques should also be able to rapidly track changes in PMD, including changes both in the DGD and the PSPs. Other desired characteristics of PMD mitigation techniques are low cost and small size to minimize the impact on existing system architectures. In addition, mitigation techniques should have a small number of feedback parameters to control [39].

In this section, we describe a PMD compensator with an arbitrarily rotatable polarization controller and a single DGD element, which can be fixed [40] or variable [41]. Figure 9 shows a schematic illustration of a single-section DGD compensator. The adjustable DGD element or birefringent element is used to minimize the impact of the fiber PMD and the polarization controller is used to adjust the direction of the polarization dispersion vector of the compensator. The expression for the polarization dispersion vector after compensation, which is equivalent to the one in [42], is given by

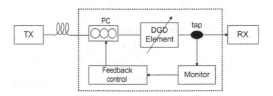

Figure 9. Schematic illustration of a single-section compensator with a monitor and a feedback element. In practical systems, the compensator will usually be part of the receiver, so that the monitor and the feedback control are integrated with the detection circuit.

$$\tau_{tot}(\omega) = \tau_c + T_c(\omega)R_{pc}\tau_f(\omega),\tag{11}$$

where τ_c is the polarization dispersion vector of the compensator, $\tau_f(\omega)$ is the polarization dispersion vector of the transmission fiber, R_{pc} is the polarization transformation in Stokes space that is produced by the polarization controller of the compensator, and $T_c(\omega)$ is the polarization transformation in Stokes space that is produced by the DGD element of the compensator. We model the polarization transformation R_{pc} as

$$R_{pc} = R_x(\phi_{pc})R_y(\psi_{pc})R_x(-\phi_{pc}).\tag{12}$$

We note that the two parameters of the polarization controller's angles in (12) are the only free parameters that a compensator with a fixed DGD element possesses, while the value of the DGD element of a variable DGD compensator is an extra free parameter that must be adjusted during the operation. In (12), the parameter ϕ_{pc} is the angle that determines the axis of polarization rotation in the y-z plane of the Poincaré sphere, while the parameter ψ_{pc} is the angle of rotation around that axis of polarization rotation. An appropriate selection of these two angles will transform an arbitrary input Stokes vector into a given output Stokes vector. While most electronic polarization controllers have two or more parameters to adjust that are different from ϕ_{pc} and ψ_{pc}, it is possible to configure them to operate in accordance to the transformation matrix R_{pc} in (12) [43].

In all the work reported in this chapter, we used the eye opening as the feedback parameter for the optimization algorithm unless otherwise stated. We defined the eye opening as the difference between the lowest mark and the highest space at the decision time in the received electrical noise-free signal. The eye-opening penalty is defined as the ratio between the back-to-back and the PMD-distorted eye opening. The back-to-back eye opening is computed when PMD is not included in the system. Since PMD causes pulse spreading in amplitude-shift keyed modulation formats, the isolated marks and spaces are the ones that suffer the highest penalty [44]. To define the decision time, we recovered the clock using an algorithm based on one described by Trischitta and Varma [45].

We simulated the 16-bit string "0100100101101101." This bit string has isolated marks and spaces, in addition to other combinations of marks and spaces. In most of other simulations in this dissertation we use pseudorandom binary sequence pattern. The receiver model consists of an Gaussian optical filter with full width at half maximum (FWHM) of 60 GHz, a square-law photodetector, and a fifth-order electrical Bessel filter with a 3 dB bandwidth of 8.6 GHz. To determine the decision time after the electronic receiver, we delayed the bit stream by half a bit slot and subtracted it from the original stream, which is then squared. As a result a strong tone is produced at 10 GHz. The decision time is set equal to the time at which the phase of this tone is equal to $\pi/2$. The goal of our study is to determine the performance limit of the compensators. In order to do that, we search for the angles ϕ_{pc} and ψ_{pc}.

of the polarization controller for which the eye opening is largest. In this case, the eye opening is our compensated feedback parameter. We therefore show the global optimum of the
compensated feedback parameter for each fiber realization.

To obtain the optimum, we start with 5 evenly spaced initial values for each of the angles
ϕ_{pc} and ψ_{pc} in the polarization transformation matrix R_{pc}, which results in 25 different initial
values. If the DGD of the compensator is adjustable, we start the optimization with the DGD
of the compensator equal to the DGD of the fiber. We then apply the conjugate gradient algorithm [46] to each of these 25 initial polarization transformations. To ensure that this procedure yields the global optimum, we studied the convergence as the number of initial
polarization transformations is increased. We examined 10^4 fiber realizations spread
throughout our phase space, and we never found more than 12 local optima in the cases that
we examined. We missed the global optimum in three of these cases because several optima
were closely clustered, but the penalty difference was small. We therefore concluded that 25
initial polarization transformations were sufficient to obtain the global optimum with sufficient accuracy for our purposes. We observed that the use of the eye opening as the objective function for the conjugate gradient algorithm produces multiple optimum values when
both the DGD and the length of the frequency derivative of the polarization dispersion vector are very large.

The performance of the compensator depends on how the DGD and the effects of the first-
and higher-order frequency derivatives of the polarization dispersion vector of the transmission fiber interact with the DGD element of the compensator to produce a residual
polarization dispersion vector and on how the signal couples with the residual principal
states of polarization over the spectrum of the channel. Therefore, the operation of single-
section PMD compensators is a compromise between reducing the DGD and setting one
principal state of polarization after compensation that is approximately co-polarized with
the signal. An expression for the pulse spreading due to PMD as a function of the polarization dispersion vector of the transmission fiber and the polarization state over the spectrum
of the signal was given in [47].

3.2. Three-section compensator

Second-order PMD has two components: Polarization chromatic dispersion (PCD) and the
principal states of polarization rotation rate (PSPRR) [35]. Let τ_1 be the polarization dispersion vector of the transmission line, and let τ_2 and τ_3 be the polarization dispersion vec-
tors of the two fixed-DGD elements of the three-section compensator. Using the
concatenation rule [42], the first- and second-order PMD vector of these three concatenated fibers are given by

$$\tau_{tot} = R_3 R_2 \tau_1 + R_3 \tau_2 + \tau_3, \tag{13}$$

$$\tau_{tot,w} = \left(\tau_3 + R_3\,\tau_2\right) \times R_3 R_2 \tau_1 q_1 + \tau_3 \times R_3 \tau_2 + R_3 R_2 \tau_{1w} q_1 + R_3 R_2 \tau_1 q_{1w}, \tag{14}$$

where R_2 and R_3 are the rotation matrices of the polarization controllers before the first and the second fixed-DGD elements of the compensator, respectively. In (14), $\tau_{1w}q_1$ and $\tau_1 q_{1w}$ are the transmission line PCD and the PSPRR components, respectively, where we express the polarization dispersion vector of the transmission fiber as $\tau_1 = \tau_1 q_1$. Here, the variable τ is the DGD and $q = \tau / |\tau|$ is the Stokes vector of one of the two orthogonal principal states of polarization. The three-section PMD compensator has two operating points [35]. For the first operating point, the term $\tau_3 \times R_3 \tau_2$ in (14) is used to cancel the PSPRR component $R_3 R_2 \tau_1 q_{1w}$, provided that we choose R_3 and R_2 so that $R_3^\dagger \tau_3 \times \tau_2$ and $R_2 \tau_1 q_{1w}$ are antiparallel, where $R_3^\dagger$ is the Hermitian conjugate of R_3. Note that with this configuration one cannot compensate for PCD.

For the second operating point, $\tau_3 \times R_3 \tau_2$ in (14) is used to compensate for PCD by choosing $R_3^\dagger \tau_3 \times \tau_2$ and $R_2 \tau_{1w} q_1$ to be antiparallel. Moreover, we can add an extra rotation to R_2 so that $\left[\left(R_3^\dagger \tau_3 + \tau_2\right) \times R_2 \tau_1 q_1\right]$ and $R_2 \tau_1 q_{1w}$ are also antiparallel. In this way, the compensator can also reduce the PSPRR term. In our simulations, we computed the reduction of the PCD and PSPRR components for the two operating points and we selected the one that presented the largest reduction of the second-order PMD. Finally, the third, variable-DGD, section of the compensator cancels the residual DGD τ_{tot} after the first two sections.

4. Simulation results and discussions

We evaluate the performance of optical fiber communication systems with and without PMD compensators using the statistical methods of importance sampling (IS) and multicanonical Monte Carlo (MMC). Both MMC and IS can be used to bias Monte Carlo simulations to the outage probability due to PMD in optical fiber communication systems with one-section and with three-section PMD compensators. When there exist a IS bias technique available, IS is more effective than MMC because each sample in IS is independent, while the samples in MMC slowly become uncorrelated. However, the effectiveness of MMC can be comparable or even exceed that of IS in the cases in which there isn't a high correlation between the parameters that are biased in IS and the parameter of interest. This is the case of optical communication systems with PMD compensation, in which IS has to exploit a vast region of the probability space that does not contribute to the events of interest.

In Fig.10, we show the pdf of the eye-opening penalty for a system with 30 ps mean DGD and a single-section compensator. We compute the pdf using IS in which only the DGD is biased, and we also compute the pdf using IS in which both the first- and the second-order PMD are biased. We observed that it is not sufficient to only bias the DGD in order to accurately calculate the compensated penalty and its pdf. This approach can only be used in systems where the DGD is the dominant source of penalties, which is the case in uncompensated systems and in systems with limited PMD compensation.

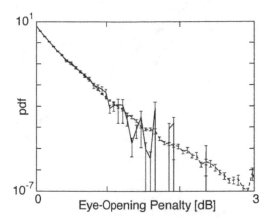

Figure 10. PDF of the eye-opening penalty for a system with a mean DGD of 30 ps and a single-section compensator. (i) Solid line: results using IS in which only the DGD is biased. (ii) Dashed line: results using IS in which both first-and second-order PMD are biased. The confidence interval is shown with error bars.

In Fig. 11, we show the outage probability as a function of the eye-opening penalty. We apply the MMC algorithm to compute PMD-induced penalties in a 10 Gbit/s NRZ system using 50 MMC iterations with 2,000 samples each. The results obtained using the samples in the final iteration of the MMC simulation (dashed and solid lines) are in excellent agreement with the ones obtained using importance sampling (open circles and squares). Here we used the results computed with importance sampling to validate the results obtained with MMC. The use of importance sampling to compute penalties in PMD single-section compensated systems was already validated with a large number of standard Monte Carlo simulations by Lima Jr. *et al.* [36], [14]. Therefore, the results computed with importance sampling can be used to validate the results computed with MMC. Our goal here is to show the applicability of MMC to accurately compute PMD-induced penalties in uncompensated and single-section PMD compensated systems.

In Fig. 12, we show contours (dotted lines) of the joint pdf of the magnitude of the uncompensated normalized first- and second-order PMD, $|\tau|$ and $|\tau_\omega|$, computed using importance sampling, as in [48]. We also show contours for the eye-opening penalty (solid lines) of an uncompensated system with a mean DGD, $\langle |\tau| \rangle$, of 15 ps. The penalty contours were produced using the same samples we generated using the MMC method in the computation of the outage probability shown in Fig. 11. The MMC method automatically placed its samples in the regions of the $|\tau|-|\tau_\omega|$ plane that corresponds to the large DGD values that have the highest probability of occurrence, which is the region that is the dominant source of penalties in uncompensated systems.

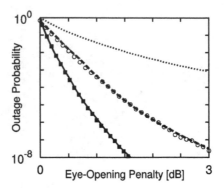

Figure 11. Outage probability as a function of the eye-opening penalty. (i) Dotted line: Uncompensated system with a mean DGD of 30 ps. (ii) Dashed line and (iii) Open circles: Results for a variable-DGD compensator, obtained using MMC and IS, respectively, for a system with mean DGD of 30 ps. (iv) Solid line and (v) Squares: Results for an uncompensated system with mean DGD of 15 ps, obtained using MMC and IS, respectively.

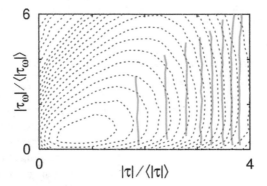

Figure 12. Penalty curves computed with MMC for an uncompensated system. Uncompensated system with a mean DGD of 15 ps. The dotted lines show the contour plots of the joint pdf of the normalized $|\tau|$ and $|\tau_\omega|$, obtained using IS. The solid lines show the average eye-opening penalty given a value of $|\tau|$ and $|\tau_\omega|$, obtained using MMC. The contours of joint pdf from the bottom to the top of the plot, are at 3×10^{-n}, and, $n=1, ..., 7$ and 10^{-m}, $m=1, ...,11$. The penalty contours in dB from the left to the right of the plot, are at 0.2, 0.4, 0.6, 0.8, 1.0, 1.2, 1.4, 1.6.

In Fig. 13, we show similar results for a system with $\langle |\tau| \rangle$=30 ps and a variable-DGD compensator that was programmed to minimize the residual DGD at the central frequency of the channel after compensation. In contrast to Fig.12, the MMC method automatically placed its samples in the regions of the $|\tau| - |\tau_\omega|$ plane where $|\tau_\omega|$ is large and the DGD is close to its average, corresponding to the region in the plane that is the dominant source of penalties in this compensated system. These results agree with the fact that the contour plots in the region dominating the penalty are approximately parallel to the DGD axis, indicating that the penalty is nearly independent of DGD. In Figs. 12 and 13, the samples obtained using the MMC method are automatically biased towards the specific region of the $|\tau| - |\tau_\omega|$ plane that dominates the penalty, *i.e.*, the region where the corresponding penalty level curve intersects the contour of the joint pdf of $|\tau|$ and $|\tau_\omega|$ with the highest probability. We did not compute the confidence interval for the results showed in this section.

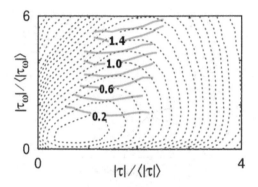

Figure 13. Same set of curves of Fig. 12 for a compensated system with a variable-DGD compensator. The penalty contours in dB from the bottom to the top of the plot, are at 0.2, 0.4, 0.6, 0.8, 1.0, 1.2, 1.4, 1.6.

In the following results, we evaluated the performance of a single-section and a three-section PMD compensator in a 10 Gbit/s nonreturn-to-zero system with a mean DGD of 30 ps. We used perfectly rectangular pulses filtered by a Gaussian shape filter that produces a rise time of 30 ps. We simulated a string with 8 bits generated using a pseudorandom binary sequence pattern. We modeled the fiber using the coarse step method with 80 birefringent fiber sections, which reproduces first- and higher-order PMD distortions within the probability range of interest [14]. The results of our simulations can also be applied to 40 Gbit/s systems by scaling down all time quantities by a factor of four. As in previous results, we used the eye opening for performance evaluation. The three-section compensator has two fixed-DGD elements of 45 ps and one variable-DGD element. The results that we present in this section were obtained using 30 MMC iterations with 8,000 samples each and using importance sampling with a total of 2.4×10^5 samples. We estimated the errors in MMC using the transition matrix method that we described in Section 2.2, while we estimated the errors in importance sampling as in [25].

In Fig. 14, we show the outage probability for a 1-dB penalty as function of the DGD element (τ_c) for a system with the three-section compensator that we used. We observed that there is an optimum value for τ_c that minimizes the outage probability, which is close to 45 ps. We set the values for the two fixed-DGD elements of the three-section PMD compensator that we used to this optimum value. The reason why the outage probability rises when τ_c

becomes larger than this optimum is because large values of τ_c add unacceptable penalties to fiber realizations with relatively small second-order PMD values that could be adequately compensated at lower values of τ_c. We also observed that there is a relatively small dependence of the outage probability on τ_c. That is because the third, variable-DGD section of the compensator cancels the residual DGD after the first two sections, which significantly mitigates the penalty regardless of the value of τ_c.

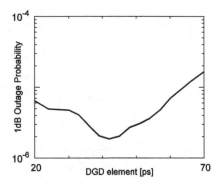

Figure 14. Outage probability for a 1-dB penalty as function of the DGD element (τ_c) of the three-section compensator for a system with mean DGD of 30 ps.

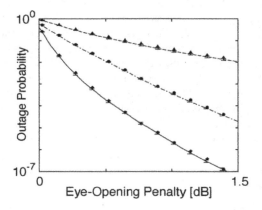

Figure 15. Outage probability as a function of the eye-opening penalty for a system with mean DGD of 30 ps. (i) Dashed line (MMC) and triangles (IS): Uncompensated system. (ii) Dot-dashed line (MMC) and circles (IS): System with a single-section compensator. (iii) Solid line (MMC) and diamonds (IS): System with a three-section compensator. The error bars show the confidence interval for the MMC results.

In Fig. 15, we plot the outage probability $(\hat{P}_{out})$ as a function of the eye-opening penalty for the compensators that we studied. The histogram of the penalty was divided into 34 evenly spaced bins in the range –0.1 and 2 dB, even though we show results from 0 to 1.5 dB of penalty. The maximum relative error $(\hat{\sigma}_{\hat{P}_{out}} / \hat{P}_{out})$ for the curves computed with MMC shown in this plot equals 0.13. The relative error for the curves computed with importance sampling is smaller than with MMC, and is not shown in the plot. The maximum relative error for the curves computed with importance sampling equals 0.1. The results obtained using MMC (solid lines) are in agreement with the ones obtained using importance sampling (symbols). The agreement between the MMC and importance sampling results was expected for the case that we used a single-section compensator, since this type of compensator can only compensate for first-order PMD [6], so that the dominant source of penalty after compensation is the second-order PMD of the transmission line. Hence, it is expected that MMC and importance sampling give similar results. We also observed good agreement between the MMC and importance sampling results for the three-section compensator. This level of agreement indicates that three-section compensators that compensate for the first two orders of the Taylor expansion of the transmission line PMD produce residual third and higher orders of PMD that are significantly correlated with the first- and second-order PMD of the transmission line. That is why the use of importance sampling to bias first- and second-order PMD is sufficient to accurately compute the outage probability in systems where the first two orders of PMD of the transmission line are compensated.

Significantly, we observed that the performance improvement with the addition of two sections, from the single-section compensator to the three-section compensator, is not as large as the improvement in the performance when one section is added, from the uncompensated to the single-section compensator. The diminishing returns that we observed for increased compensator complexity is consistent with the existence of correlations between the residual higher orders of PMD after compensation and the first two orders of PMD of the transmission line that are compensated by the three-section compensator.

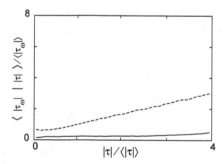

Figure 16. Conditional expectation of the magnitude of the normalized second-order PMD, $\left| \tau_\omega \right|$, given a value of the DGD of the transmission line, $| \tau |$. Conditional expectation before (dashed) and after (solid) the three-section compensator.

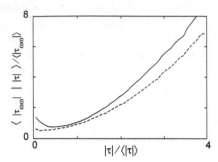

Figure 17. Conditional expectation of the magnitude of the normalized third-order PMD, $|\tau_{\omega\omega}|$, given a value of the DGD of the transmission line, $|\tau|$. Conditional expectation before (dashed) and after (solid) the three-section compensator.

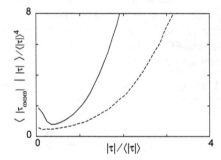

Figure 18. Conditional expectation of the magnitude of the normalized fourth-order PMD, $|\tau_{\omega\omega\omega}|$, given a value of the DGD of the transmission line, $|\tau|$. Conditional expectation before (dashed) and after (solid) the three-section compensator.

Figures 16–18 quantify the correlation between the lower and higher orders of PMD. In Fig. 16, we show the conditional expectation of the magnitude of second-order of PMD both before and after the three-section compensator given a value of the DGD of the transmission line. In these figures, the DGD $|\tau|$ is normalized by the mean DGD $\langle|\tau|\rangle$ and $|\tau_{\omega}|$ is normalized by $\langle|\tau_{\omega}|\rangle$ to obtain results that are independent of the mean DGD and of the mean of the magnitude of second-order PMD. We observed a large correlation between $|\tau|$ and $|\tau_{\omega}|$ before compensation, while after compensation $|\tau_{\omega}|$ is significantly reduced and is less correlated with the DGD, demonstrating the effectiveness of the three-section compensator in compensating for second-order PMD. In Figs. 17 and 18, we show the conditional expectation of the magnitude of the third-order PMD and of the fourth-order PMD, respectively, before and after the three-section compensator, given a value of the DGD of the transmission line. In both cases, we observed a high correlation of the third- and the fourth-order PMD with the DGD before and after compensation. In addition, we observed a significant increase of these higher-order PMD components after compensation, which leads to a residual penalty after compensation that is correlated to the original first- and second-order PMD.

In Fig. 19, we show contour plots of the conditional expectation of the penalty with respect to the first- and second-order PMD for a system with a three-section PMD compensator [35]. These results show that the residual penalty after compensation is significantly correlated with the first- and second-order PMD. The correlation between the higher orders of PMD with the DGD that we show in Figs. 16–18 can be estimated from the concatenation rule [42], which explicitly indicates a dependence of the higher-order PMD components on the lower order components. The increase in these higher-order components after compensation is also due to our choice of the operating point of this compensator, which is set to compensate only for first- and second-order PMD, regardless of the higher-order PMD components. It is possible that this three-section PMD compensator would perform better if all 7 parameters of the compensator are adjusted to achieve the global penalty minimum. However, finding this global optimum is unpractical due to the large number of local optima in such a multi-dimensional optimization space, as we found in our investigation of single-section PMD compensators [14]. On the other hand, the compensation of first- and second-order PMD using the three-section compensator that we studied here, which was proposed by Zheng, *et al.* [35], can be implemented in practice.

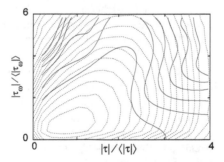

Figure 19. Three-section compensated system. The dotted lines are contour plots of the joint pdf of the normalized $|\tau|$ and $|\tau_\omega|$ from the bottom to the top of the plot, are at 3×10^{-n}, with $n=1,\cdots,7$ and 10^{-m}, with $m=1,\cdots,11$. The solid lines are contour plots of the conditional expectation of the eye-opening penalty in dB from the bottom to the top of the plot, are at 0.1, 0.2, 0.3, 0.4, 0.5, 0.6.

In this Section, we showed that both multiple importance sampling and MMC can be used with all the compensators that we investigated to reduce the computation time for the outage probability due to PMD in optical fiber communication systems. Importance sampling in which both the first- and second-order PMD are biased can be used to efficiently compute the outage probability even with a three-section PMD compensator in which both first- and second-order PMD are compensated, which is consistent with the existence of a large correlation between first- and second-order PMD of the transmission line and higher orders of PMD after compensation. We directly verified the existence of these correlations. In contrast to what we presented in Fig.11, where importance sampling was used to validate the results with MMC, in the resulted subsequently presented, we used MMC to validate the results obtained with importance sampling. We used MMC to validate the results obtained with importance sampling because MMC can be used to compute penalties induced by all orders of PMD and not just penalties correlated to first- and second-order PMD as is the case with the importance sampling method. We showed that MMC yields the same results as importance sampling, within the statistical errors of both methods. Finally, we showed that the three-section compensator offers less than twice the advantage (in dB) of single-section compensators. We attribute the diminishing returns with increased complexity to the existence of correlations between the first two orders of PMD prior to compensation and higher orders of PMD after compensation.

5. Conclusions

In this chapter, we used MMC and IS in which both the first- and second-order PMD are biased to investigate the performance of single-section and three-section PMD compensators. We showed that both methods are effective to compute outage probabilities for the optical fiber communication systems that we studied with and without PMD compensators. The comparison of importance sampling to the MMC method not only allowed us to mutually validate both calculations, but yielded insights that were not obtained from either method alone. The development of IS requires some *a priori* knowledge of how to bias a given parameter in the simulations. In this particular problem, the parameter of interest is the penalty. However, to date there is no IS method that directly biases the penalty. Instead of directly biasing the penalty, one has to rely on the correlation of the first-and second-order PMD with the penalty, which may not hold in all compensated systems. In contrast to IS, MMC does not require *a priori* knowledge of which rare events contribute significantly to the penalty distribution function in the tails, since the bias is done automatically in MMC. Because the samples in IS are independent, IS converges more rapidly than MMC when the biased quantity is highly correlated to the parameter of interest. However this is not always the case. The applicability of IS to model a system with a three-section PMD compensator, in which both first- and second-order components of the Taylor's expansion of PMD in the frequency domain are compensated, is consistent with the existence of a large correlation between first- and second-order PMD components of the transmission line and the higher orders of PMD after compensation. Thus, even when the first two orders of PMD are compensated, these quantities prior to compensation still remain highly correlated with the residual penalty.

It is essential to carefully monitor statistical errors when carrying out Monte Carlo simulations in order to verify the accuracy of the results. Effective procedures for calculating the statistical errors in standard Monte Carlo simulations are well known and are easily implemented. Moreover, in this case, each sample is independently drawn, and the errors in each bin of the histogram will also be independent. Hence, the smoothness of the histogram is often a good indication that the errors are acceptably low. While calculating the statistical errors with importance sampling is more complicated, analytical formulae have been successfully implemented. By contrast, calculating statistical errors using MMC is not trivial. MMC generates correlated samples, so that standard error estimation techniques cannot be applied. To enable the estimate of the statistical errors in the calculations using MMC we developed a method that we refer to as the MMC transition matrix method. The method is based on the calculation of a transition matrix with a standard MMC simulations and the use of this transition matrix to draw a large number of independent samples.

Author details

Aurenice M. Oliveira[1] and Ivan T. Lima Jr.[2]

1 Michigan Technological University, U.S.A

2 North Dakota State University, U.S.A

References

[1] B. Huttner, C. Geiser, and N. Gisin, "Polarization-induced distortions in optical fiber networks with polarization-mode dispersion and polarization-dependent losses," *IEEE J. Selec. Topics Quantum Electron*, vol. 6, pp. 317-329, 2000.

[2] G. Biondini, W. L. Kath, and C. R. Menyuk, "Importance sampling for polarization-mode dispersion," *IEEE Photon. Technol. Lett*, vol. 14, pp. 310-312, 2002.

[3] S. L. Fogal, G. Biondini, and W. Kath, "Multiple importance sampling for first- and second-order polarization-mode dispresion," *IEEE Photon. Technol Lett*, vol. 14, pp. 1273-1275, 2002.

[4] B. A. Berg and T. Neuhaus, "The multicanonical ensemble: a new approach to simulate first-order phase transitions, "*Phys. Rev. Lett*, vol. 68, pp. 9-12, 1992.

[5] A. M. Oliveira, I. T. LimaJr, C. R. Menyuk, G. Biondini, B. Marks, and W. Kath, "Statistical analysis of the performance of PMD compensators using multiple importance sampling," *IEEE Photon. Technol. Lett.*, vol. 15, pp. 1716-1718, 2003.

[6] A. M. Oliveira, I. T. LimaJr, J. Zweck, and C. R. Menyuk, "Efficient computation of PMD-induced penalties using multicanonical Monte Carlo simulations," in *Proceedings ECOC 2003*, pp. 538-539.

[7] A. M. Oliveira, C. Menyuk, and I. LimaJr, "Comparison of Two Biasing Monte Carlo Methods for Calculating Outage Probabilities in Systems with Multisection PMD Compensators," *IEEE Photon. Technol. Lett.*, vol. 17, pp. 2580-2582, 2005.

[8] A. M. Oliveira, I. T. LimaJr, C. R. Menyuk, and J. Zweck, "Performance evaluation of single-section and three-section PMD compensators using extended Monte Carlo methods," in *Proceedings OFC 2005*.

[9] A. Bononi, L. Rusch, A. Ghazisaeidi, F. Vacondio, and N. Rossi, "A Fresh Look at Multicanonical Monte Carlo from a Telecom Perspective," in *IEEE Globecom 2009*, 2009.

[10] A. Bononi and L. Rusch, "Multicanonical Monte Carlo for Simulation of Optical Links," in *Impact of Nonlinearities on Fiber Optic Communications*, S. Kumar, Ed., ed: Springer Science Business Media, LLC, 2011, pp. 373-413.

[11] R. Holzlohner and C. R. Menyuk, "Use of multicanonical Monte Carlo simulations to obtain accurate bit error rates in optical communication systems," *Optics Letters*, vol. 28, pp. 1894-1896, 2003.

[12] A. Ghazisaeidi, F. Vacondio, A. Bononi, and L. Rusch, "SOA Intensity Noise Suppression in Spectrum Sliced Systems: A Multicanonical Monte Carlo Simulator of Extremely Low BER," *Journal of Lighwave Technologies*, vol. 27, pp. 2667-2677, 2009.

[13] L. Gerardi, M. Secondini, and E. Forestieri, "Performance Evaluation of WDM Systems Through Multicanonical Monte Carlo Simulations," *Journal of Lighwave Technologies*, vol. 29, pp. 871-879, 2011.

[14] I. T. LimaJr, A. M. Oliveira, G. Biondini, C. R. Menyuk, and W. L. Kath, "A comparative study of single-section polarization-mode dispersion compensators," *J. Lightwave Technol.*, vol. 22, pp. 1023-1032, 2004.

[15] A. M. Oliveira, I. T. LimaJr, C. R. Menyuk, and J. Zweck, "Error estimation in multicanonical Monte Carlo simulations with applications to polarization-mode-dispersion emulators," *Journal of Lighwave Technologies*, vol. 23, pp. 3781-3789, 2005.

[16] B. A. Berg and T. Neuhaus, "The multicanonical ensemble: a new approach to simulate first-order phase transitions," *Phys. Rev. Lett.*, vol. 68, pp. 9-12, 1992.

[17] B. A. Berg, "Multicanonical simulations step by step," *Comp. Phys. Commum.*, vol. 153, pp. 397-407, 2003.

[18] N. Metropolis, A. W. Rosenbluth, M. N. Rosenbluth, A. H. Teller, and E. Teller, "Equation of state calculations by fast computing machines," *J. Chem. Phys.*, vol. 21, pp. 1087-1092, 1953.

[19] D. Yevick, "The accuracy of multicanonical system models," *IEEE Photon. Technol. Lett.*, vol. 15, pp. 224-226, 2003.

Multicanonical Monte Carlo Method Applied to the Investigation of Polarization Effects in
Optical Fiber Communication Systems

105

[20] D. Marcuse, C. R. Menyuk, and P. K. A. Wai, "Application of the Manakov-PMD equation to studies of signal propagation in optical fibers with randomly varying birefringence," *J. Lightwave Technol.*, vol. 15, pp. 1735-1746, 1997.

[21] M. Karlsson, "Probability density functions of the differential group delay in optical fiber communication systems," *J. Lightwave Technol.*, vol. 19, pp. 324-331, 2001.

[22] M. H. Kalos and P. A. Whitlock, *Monte Carlo Methods*: John Wiley and Sons, 1986.

[23] D. Yevick, "Multicanonical communication system modeling---application to PMD statistics," *IEEE Photon. Technol. Lett.*, vol. 14, pp. 1512-1514, 2002.

[24] S. Kachigan, *Multivariate Statistical Analysis: A Conceptual Introduction*: Radius Press, 1991.

[25] A. M. Oliveira, I. T. LimaJr, C. R. Menyuk, G. Biondini, B. S. Marks, and W. L. Kath, "Statistical analysis of the performance of PMD compensators using multiple importance sampling," *IEEE Photon. Technol. Lett.*, vol. 15, pp. 1716-1718, 2003.

[26] R. Walpole and R. Myers, *Probability and Statistics for Engineers and Scientists*: Macmillian, 1993.

[27] B. Efron, "Bootstrap methods: another look at the Jackknife," *The Annals of Statistics*, vol. 7, pp. 1-26, 1979.

[28] I. T. LimaJr, R. Khosravani, P. Ebrahimi, E. Ibragimov, A. E. Willner, and C. R. Menyuk, "Comparison of polarization mode dispersion emulators," *J. Lightwave Technol.*, vol. 19, pp. 1872-1881, 2001.

[29] R. Holzlöhner and C. R. Menyuk, "The use of multicanonical Monte Carlo simulations to obtain accurate bit error rates in optical communications systems," *Opt. Lett.*, vol. 28, pp. 1894-1897, 2003.

[30] K. Singh, "On the asymptotic accuracy of Efron's Bootstrap," *The Annals of Statistics*, vol. 9, pp. 1187-1195, 1981.

[31] B. Efron and R. Tibshirani, *An Introduction to the Bootstrap*: Chapman and Hall, 1993.

[32] J. Booth and P. Hall, "Monte Carlo approximation and the Iterated Bootstrap," *Biometrika*, vol. 81, pp. 331-340, 1994.

[33] R. Holzlöhner, A. Mahadevan, C. R. Menyuk, J. M. Morris, and J. Zweck, "Evaluation of the very low BER of FEC codes using dual adaptive importance sampling," *Comm. Lett.*, vol. 9, pp. 163-165, 2005.

[34] R. Noé, D. Sandel, M. Yoshida-Dierolf, S. Hinz, V. Mirvoda, A. Schöpflin, C. Glingener, E. Gottwald, C. Scheerer, G. Fisher, T. Weyrauch, and W. Haase, "Polarization mode dispersion compensation at 10, 20, and 40 Gb/s with various optical equalizers," *J. Lightwave Technol.*, vol. 17, pp. 1602-1616, 1999.

[35] Y. Zheng, B. Yang, and X. Zhang, "Three-stage polarization mode dispersion compensator capable of compensating second-order polarization model dispersion," *IEEE Photon. Technol. Lett.*, vol. 14, pp. 1412-1414, 2002.

[36] I. T. LimaJr, G. Biondini, B. S. Marks, W. L. Kath, and C. R. Menyuk, "Analysis of PMD compensators with fixed DGD using importance sampling," *IEEE Photon. Technol. Lett.*, vol. 14, pp. 627-629, 2002.

[37] H. Sunnerud, C. Xie, M. Karlsson, R. Samuelssonm, and P. A. Andrekson, "A comparison between different PMD compensation techniques," *J. Lightwave Technol.*, vol. 20, pp. 368-378, 2002.

[38] A. M. Oliveira, I. T. LimaJr, T. Adalı, and C. R. Menyuk, "A Novel polarization diversity receiver for PMD mitigation," *IEEE Photon. Technol. Lett.*, vol. 14, pp. 465-467, 2002.

[39] I. P. Kaminow and T. Li, *Optical Fiber Telecommunications* vol. IV-B: Academic, 2002.

[40] T. Takahashi, T. Imai, and M. Aiki, "Automatic compensation technique for timewise fluctuation polarization mode dispersion in in-line amplifier systems," *Electron. Lett.*, vol. 30, pp. 348-349, 1994.

[41] F. Heismann, "Automatic compensation of first-order polarization mode dispersion in a 10 Gbit/s Transmission System," in *Proceedings ECOC 1998*, pp. 329-330.

[42] J. P. Gordon and H. Kogelnik, "PMD fundamentals: Polarization mode dispersion in optical fibers," *Proc. Nat. Acad. Sci.*, vol. 97, pp. 4541-4550, 2000.

[43] I. T. LimaJr, A. M. Oliveira, and C. R. Menyuk, "A comparative study of single-section polarization-mode dispersion compensators," *Journal of Lighwave Technologies*, vol. 22, pp. 1023-1032, 2004.

[44] H. Bülow, "System outage probability due to first- and second-order PMD," *IEEE Photon. Technol. Lett.*, vol. 10, pp. 696-698, 1998.

[45] P. R. Trischitta and E. L. Varma, *Jitter in Digital Transmission Systems*: Artech House, 1989.

[46] E. Polak, *Computational Methods in Optimization*: Academic Press, 1971.

[47] M. Karlsson, "Polarization mode dispersion-induced pulse broadening in optical fibers," *Optics Lett.*, vol. 23, pp. 688-690, 1998.

[48] A. M. Oliveira, I. T. LimaJr, B. S. Marks, C. R. Menyuk, and W. L. Kath, "Performance analysis of single-section PMD compensators using multiple importance sampling," in *Proceedings OFC 2003*, pp. 419-421.

Chapter 5

Multimode Graded-Index Optical Fibers for Next-Generation Broadband Access

David R. Sánchez Montero and
Carmen Vázquez García

Additional information is available at the end of the chapter

1. Introduction

Growing research interests are focused on the high-speed telecommunications and data communications networks with increasing demand for accessing to the Internet even from home. For instance, in Nov 2011 Strategy Analytics forecasted that there would be more than 807 million broadband fixed line subscriptions worldwide in 2017, based on a figure of 578 million at the end of 2011 showing a cumulative annual growth rate of around 8 percent [1]. This increasing demand for high-speed information transmission over the last two decades has been driven by the huge successes during the last decade of new multimedia services, commonly referred as Next-Generation Access (NGA) services, such as Internet Protocol Television (IPTV) or Video on Demand (VoD), as well as an increased data traffic driven by High-Definition TV (HDTV) and Peer-to-Peer (P2P) applications which have changed people's habits and their demands for service delivery. Consequently, consumer adoption of broadband access to facilitate use of the Internet for knowledge, commerce and, obviously, entertainment is contingent with the increment of the optical broadband access network capacity, which should extent into the customer's premises up to the terminals. Thus, steady increases in bandwidth requirements of access networks and local area networks (LANs) have created a need for short-reach and medium-reach links supporting data rates of Gbps (such as Gigabit Ethernet, GbE), 10Gbps (such as 10-Gigabit Ethernet, 10GbE) and even higher (such as 40- and 100- Gigabit Ethernet standards, namely 40GbE and 100GbE respectively, which started in November 2007 and have been very ratified in June 2010). Detailed studies [2, 3] have defined the required bit-rates to be transmitted to the customer's premises for different profiles for the traffic flows, reaching a total future-proof very-high-bit-rate link in the order of 2Gbps per user. It is estimated that end-user access bandwidths could reach 1 Gbps by 2015, and 10 Gbps by 2020.

Related to this latter premise, a growing number of service providers are turning to solutions capable of exploiting the full potential of optical fiber for service delivery, being the copper based x-Digital Subscriber Line (xDSL) infrastructure progressively replaced by a fiber-based outside plant with thousands of optical ports and optical fiber branches towards residential and business users, constituting the core of the FTTx (Fiber to the Home/Node/Curb/Business) deployments, see Fig. 1(a). These include passive optical networks (PONs), whose standardization has accelerated product availability and deployment. The ongoing evolution to deliver Gigabit per second Ethernet and the growing trend to migrate to Wavelength Division Multiplexing (WDM) schemes have benefited significantly from the Coarse WDM (CWDM) and Dense WDM (DWDM) optoelectronics technologies, as they provide a more efficient way to deliver traffic to Customer Premises Equipment (CPE) devices. These systems, commonly referred to as WDM-PON, are still under standardization process and field trials and are the basis of the so-called next-generation broadband optical access networks to prepare for the future upgrade of the FTTx systems currently being deployed. A basic scheme of the WDM-PON architecture is depecited in Fig. 1(b). However, networking architectures such as PON, BPON, WDM-PON, etc. are outside the scope of this chapter. There is a widely-spread consensus concerning service providers that FTTx is the most powerful and future-proof access network architecture for providing broadband services to residential users.

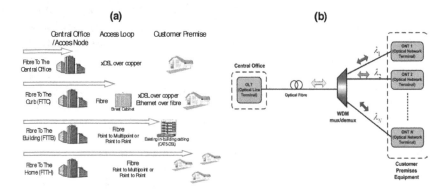

Figure 1. (a) Different FTTx network deployments. (b) Architecture of WDM-PON.

In the FTTx system concepts deployed up to now, singlemode optical fiber (SMF) is used, which has a tremendous bandwidth and thus a huge transport capacity for many services such as the ITU G.983.x ATM[1]-PON system. Research is ongoing to further extend the capabilities of shared SMF access networks. The installation of SMF has now conquered the core and

---

1 ATM: Asynchronous Transfer Mode.

metropolitan area networks and is subsequently penetrating into the access networks. How-ever, it requires great care, delicate high-precision equipment, and highly-skilled personnel, being mainly deployed for long-haul fiber optic communications, constituting the so-called Optical Distribution Network (ODN) and the core telecommunication network of the next generation of optical broadband access networks. Nevertheless, as it comes closer to the end user and his residential area, the costs of installing and maintaining the fiber network become a driving factor, which seriously hampers the large-scale introduction of FTTx.

Also inside the customer's premises, there is a growing need for convergence of the multi-tude of communication networks. Presently, Unshielded Twisted copper Pair (UTP) cables are used for voice telephony, cat-5 UTP cables for high-speed data, coaxial cables for CATV[2] and FM[3] radio signals distribution, wireless Local Area Network (LAN) for high-speed data, FireWire for high-speed short-range signals, and also Power Line Communications (PLC) technology for control signals and lower-speed data. These different networks are each dedicated and optimised for a particular set of services, which also put different Quality-of-Service (QoS) demands, and suffer from serious shortcomings when they are considered to serve the increasing demand for broadband services. Also no cooperation between the net-works exists. A common infrastructure that is able to carry all the service types would alle-viate these problems. It is therefore not easy to upgrade services, to introduce new ones, nor to create links between services (e.g., between video and data). By establishing a common broadband in-house network infrastructure, in which a variety of services can be integrated, however, these difficulties can be surmounted. The transmission media used at present are not suited for provisioning high-bandwidth services at low cost. For instance, today's wiring in LANs is based mainly on copper cables (twisted pair or coaxial) and silica (glass) fiber of two kinds: singlemode optical fiber (SMF) and multimode optical fiber (MMF). Copper based technologies suffer strong susceptibility to electromagnetic interferences and have limited capacity for digital transmission as well as the presence of crosstalk. Compared to these copper based technologies, optical fiber has smaller volume, it is less bulky and has a smaller weight. In comparison with data transmission capability, optical fiber offers higher bandwidth at longer transmission distances.

On the other hand, optical fiber is extensively used for long-distance data transmission and it represents an alternative for transmission at the customer's premises as well. Optical fiber connections offer complete immunity to EMI and present increase security, since it is very difficult to intercept signals transmitted through the fiber. Moreover, optical communication systems based on silica optical fibers allow communication signals to be transmitted not on-ly over long distances with low attenuation but also at extremely high data rates, or band-width capacity. In SMF systems, this capability arises from the propagation of a single optical mode in the low-loss windows of silica located at the near-infrared wavelengths of 1.3μm, and 1.55μm. Furthermore, since the introduction of Erbium-Doped Fiber Amplifiers (EDFAs), the last decade has witnessed the emergence of SMF as the standard data trans-

---

2 CATV: Community Antena TeleVision.

3 FM: Frequency Modulation.

mission medium for wide area networks (WANs), especially in terrestrial and transoceanic communication backbones. The success of the SMF in long-haul communication backbones has given rise to the concept of optical networking, which is a central theme with currently driving research and development activities in the field of photonics. The main objective is to integrate voice, video, and data streams over all-optical systems as communication signals make their way from WANs down to the end user by Fiber-To-The-x (FTTx), offices, and in-homes.

Although conventional SMF solutions have the potential of achieving very large bandwidths, they suffer from high connections costs compared to copper or wireless solutions. For this reason, SMF has not been widely adopted by the end user (premises) where most of the interconnections are needed and less cost-sharing between users is obtainable. The underlying factor is the fact that the SMF core is typically only a few micrometers in diameter with the requirement of precise connecting, delicate installation and handling. Yet as the optical network gets closer to the end user, the system is characterized by numerous connections, splices, and couplings that make the use of thin SMF impractical. An alternative technology is then the use of conventional silica-based multimode optical fiber (MMF) with larger core diameters. This fact allows for easier light coupling from an optical source, large tolerance on axial misalignments, which results in cheaper connectors and associated equipment, as well as less requirements on the skills of the installation personnel. However, the use of MMFs is at a cost of a bandwidth penalty with regards to their SMF counterparts, mainly due to the introduction of modal dispersion. This is the reason why MMF is commonly applied to short-reach and medium-reach applications due to its low intrinsic attenuation despite its limited bandwidth. In particular, in the access network, the use of MMF may yield a considerable reduction of installation costs although the bandwidth-times length product of SMF is significantly higher than that of MMF. As in the access network, the fiber link lengths are less than 10km, however, the bandwidth of presently commercially available silica MMFs is quite sufficient.

On the other hand, compared to multimode silica optical fiber, polymer optical fiber (POF) offers several advantages over conventional multimode optical fiber over short distances (ranging from 100m to 1000m) such as the even potential lower cost associated with its easiness of installation, splicing and connecting. This is due to the fact that POF is more flexible and ductile [4], making it easier to handle. Consequently, POF termination can be realized faster and cheaper than in the case of silica MMF. This POF technology could be used for data transmission in many applications areas ranging like in-home, fiber to the building, wireless LAN backbone or office LAN among others. In addition, improvement in the bandwidth of POF fiber can be obtained by grading the refractive index, thus introducing the so-called Graded-Index POFs (GIPOFs). Although by grading the index profile significantly enhanced characteristics have been obtained, the bandwidth and attenuation still limit the transmission distances and capacity. Reduction of loss has been achieved by using amorphous perfluorinated polymers for the core material. This new type of POF has been named perfluorinated GIPOF (PF GIPOF). This new fiber with low attenuation and large bandwidth has opened the way for high capacity transmission over POF based systems. In addition to, as PF GIPOF has a relative low loss wavelength region ranging from 650nm to 1300nm (even theoretically in the third transmission window), it allows for WDM transmission of several data channels. However, attenuation and bandwidth characteristics of the current state-of-the-art PF GIPOF are not at par with those of standard silica SMFs, but they still are superior to those of copper based technologies. Nevertheless, although these losses are coming down steadily due to ongoing improvements in the production processes of this still young technology, the higher than silica attenuation inhibits their use in relative long link applications, being mainly driven for covering in-building optical networks link lengths for in-building/home optical networks (with link lengths less than 1 km), and thus the loss per unit length is of less importance. It should be noted that available light sources for silica fiber based systems can be used with PF GIPOF systems. The same is true of connectors as in the case of Gigabit Ethernet equipment.

Therefore, it can be stated that polymer optical fiber technology has reached a level of development where it can successfully replace copper based technology and silica MMF for data transmission in short distance link applications such as in the office, in-home and LAN scenarios. Moreover, PF GIPOF is forecasted to be able to support bit-rate distance products in the order of 10Gbps km [5]. Short distance communications system like in-home network and office LANs represent a unique opportunity for deployment of PF GIPOF based systems for broadband applications. We can conclude that PF GIPOF technology is experiencing rapid development towards a mature solution for data transmission at short haul communications. The challenge remains in bringing this POF technology (transceiver, connectors,...) to a competitive price and performance level at the customer's premises.

Nevertheless, the potentials of these multimode fibers, both silica- and polymer-based, to support broadband radio-frequency, microwave and even millimetre wave transmission over short- and medium-reach distances are yet to be fully known. The belief is that a better understanding of the factors that affect the fiber bandwidth will prove very useful in increasing the bandwidth of silica MMF and PF GIPOF links in real situations. In the whole fiber network society to be realized in the near future, it is said that silica-based SMF fibers for long-haul backbone will be only several percents of the total use, and the remaining more than 90% would correspond to all-optical networks covering the last mile [6]. Link lengths may range from well below 1 km in LANs and residential houses, to only a few kilometres in larger building such as offices, hospitals, airport halls, etc. And it is now clear that the expected market is huge and researches and companies all over the world are competing to find a solution to this issue.

In this framework, the first part of this chapter, comprising sections 2 and 3 will briefly address the fundamentals of mutimode optical fibers as well as present transmission capacities. Like any communication channel, the multimode optical fiber also suffers from various signal distortions limiting its usefulness. The primary mechanisms contributing to the channel impairment in multimode fibers are discussed. Both silica-based MMFs and PF GIPOFs are essentially large-core optical waveguide supporting multiple transverse electromagnetic modes and they suffer from similar channel impairments. On the other hand, present capabilities of actual multimode optical fiber-based deployments are shown. In addition, different techniques reported in literature to carry microwave and millimetre-wave over optical networks, surmounting the multimode fiber bandwidth bottleneck, are also briefly described.

The second part of this chapter, which comprises sections 4, 5 and 6, respectively, focuses on the frequency response mathematical framework and the experimental results, respectively, of both types of multimode optical fibers. Some of the key factors affecting the frequency characteristics of both fiber types are addressed and studied. Theoretical simulations and measurements are shown for standard silica-based MMF as well as for PF GIPOF. Although some of these issues are interrelated, they are separately identified for clarity.

Finally, the main conclusions of this chapter are reported in Section 7.

2. Fundamentals of multimode optical fibers

Despite the above advantages, the use of multimode optical fiber has been resisted for some years by fiber-optic link designers in favour of their SMF counterparts since Epworth discovered the potentially catastrophic problem of modal noise [7]. Modal noise in laser-based MMF links has been recently more completely addressed and theoretical as well as experimental proofs have shown that long-wavelength operation of MMFs is robust to modal noise [8-10]. This explains the spectacular regain of interest for MMFs as the best solution for the cabling of the access, in-home networks and LANs. The question that needs answer now in view of increasing the usefulness of MMF concerns the improvement of their dispersion characteristics, which is related to their reduced bandwidth.

For the transmission of communication signals, attenuation and bandwidth are important parameters. Both parameters will be briefly described in the following subsections, focusing on their impact over multimode fibers. In any case, the optical signal is distorted and attenuated when it propagates over the fiber. These effects have to be modeled when describing the signal transmission. They behave quite differently in different types of fibers. Whereas signal distortions in singlemode fibers (SMFs) are primarily caused by chromatic dispersion, i.e. the different speeds of individual spectral parts, the description of dispersion in multimode fibers (MMFs) is considerably more complex. Not only does chromatic dispersion occur in them, but also has the generally much greater modal (or intermodal) dispersion.

It should be noted that, apart from attenuation, an important characteristic of an optical fiber as a transmission medium is its bandwidth. Bandwidth is a measure of the transmission capacity of a fiber data link. As multimode fibers can guide many modes having different velocities, they produce a signal response inferior to that of SMFs, being this modal dispersion effect the limiting bandwidth factor. So bandwidth and dispersion are two parameters closely related.

2.1. Attenuation

Attenuation in fiber optics, also known as transmission loss, is the reduction in the intensity of the light beam with respect to distance traveled through a transmission medium, being an

important factor limiting the transmission of a digital signal across large distances. The attenuation coefficient usually use units of dB/km through the medium due to the relatively
high quality of transparency of modern optical transmission media. Empirical research over
the years has shown that attenuation in optical fiber is caused primarily by both scattering
and absorption. However, the fundamentals of both attenuation mechanisms are outside the
scope of this chapter.

On the one hand, silica exhibits fairly good optical transmission over a wide range of wavelengths. In the near-infrared (near IR) portion of the spectrum, particularly around 1.5 μm,
silica can have extremely low absorption and scattering losses of the order of 0.2 dB/km.
Such remarkably-low losses are possible only because ultra-pure silicon is available, being
essential for manufacturing integrated circuits and discrete transistors. Nevertheless, fiber
cores are usually doped with various materials with the aim of raising the core refractive index thus achieving propagation of light inside the fiber (by means of total internal reflection
mechanisms). A high transparency in the 1.4-μm region is achieved by maintaining a low
concentration of hydroxyl groups (OH). Alternatively, a high OH concentration is better for
transmission in the ultraviolet (UV) region.

On the other hand, until recently, the only commercially available types of POF were based
on non-fluorinated polymers such as PolyMethylMethAcrylate (PMMA) (better known as
Plexiglass®), widely used as core material for graded-index fiber [11] in addition with the
utilization of several kinds of dopants. Although firstly developed PMMA-GIPOFs were
demonstrated to obtain very high transmission bandwidth compared to that of Step-Index
(SI) counterparts, the use of PMMA is not attractive due to its strong absorption driving a
serious problem in the PMMA-based POFs at the near-IR (near-infrared) to IR regions. This
is because of the large attenuation due to the high harmonic absorption loss by carbon-hy-
drogen (C-H) vibration (C-H overtone). As a result, PMMA-based POFs could only be used
at a few wavelengths in the visible portion of the spectrum, typically 530nm and 650nm,
with typical attenuations around 150dB/km at 650nm. Today, unfortunately, almost all gigabit optical sources operate in the near-infrared (typically 850nm or 1300nm), where PMMA
and similar polymers are essentially opaque. Nevertheless, in this scenario, undistorted bit
streams of 2.5Gbps over 200m of transmission length were successfully demonstrated over
PMMA-GIPOF [12].

On the other hand, it has been reported that one can eliminate this absorption loss by substi-
tuting the hydrogen atoms in the polymer molecule for heavier atoms [13]. In this case, if the
absorption loss decreases with the substitution of hydrogen for deuterium or halogen atoms
(such as fluorine), the possible distance for signal transmission would be limited by dispersion, and not by attenuation. Many polymers have been researched and reported in literature in order to improve the bandwidth performance given by the first PMMA-based
graded-index polymer optical fibers [14]. Nevertheless, today, amorphous perfluorinated
(PF) GIPOF is widely used because of its high bandwidth and low attenuation from the visible to the near IR wavelengths compared to PMMA GIPOF [15]. As a result, it is immediately compatible with gigabit transmission sources, and can be used over distances of
hundreds of meters. This fact is achieved mainly by reducing the number of carbon-hydro-

gen bonds that exist in the monomer unit by using partially fluorinated polymers. In 1998, the PF-based GIPOF had an attenuation of around 30dB/km at 1310nm. Attenuation of 15dB/km was achieved only three years after and lower and lower values of attenuation are being achieved. The theoretical limit of PF-based GIPOFs is ~0.5 dB/km at 1250-1390nm [16]. In the estimation, the attenuation factors are divided in two: material-inherent scattering loss and material-inherent absorption loss. The first factor is mainly given by the Rayleigh scattering, following the relation $\alpha_R \sim (\lambda)^{-4}$. The second factor is given by the absorption caused by molecular vibrations. A detailed explanation on the estimation processes is described [17].

2.2. Dispersion

As aforementioned, pulse broadening in MMFs is generally caused by modal dispersion and chromatic dispersion. For MMFs it is necessary to consider the factors of material, modal and profile dispersion. The latter considers the wavelength dependence on the relative refractive index difference in graded index fibers. Waveguide dispersion additionally occurs in singlemode fibers, whereas profile dispersion and modal dispersion do not.

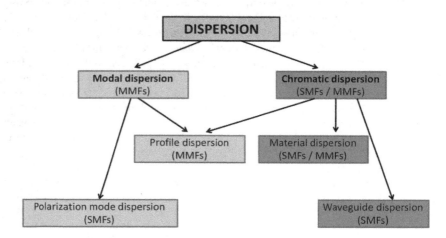

Figure 2. Dispersion mechanisms in optical fibers.

All the kinds of dispersion appearing in optical fibers are summarized in Fig. 2. The mechanisms dependent on the propagation paths are marked in blue, whereas the wavelength-dependent processes are marked in red. Those mechanisms only affecting SMFs are outside from the scope of this work so they will be avoided. For multimode fibers modal dispersion and chromatic dispersion are the relevant processes to be considered.

In a generic description, chromatic dispersion is introduced by the effect that the speed of propagation of light of different wavelengths differs resulting in a wavelength dependence of the modal group velocity. The end result is that different spectral components arrive at slightly different times, leading to a wavelength-dependent pulse spreading, i.e. dispersion. As a matter of fact, the broader the spectral width (linewidth) of the optical source the greater is the chromatic dispersion. In PF-based POFs the chromatic dispersion is much smaller than in silica MMF for wavelengths up to 1100nm. For wavelengths above 1100nm, the dispersion of the PF-based GIPOF retains and the dispersion of silica MMF increases. The expression of such dispersion is given by:

$$\Delta t_{chrom} = D(\lambda) \cdot \Delta\lambda \cdot L \ ; \ D(\lambda) = -\frac{\lambda}{c} \cdot \frac{d^2 n(\lambda)}{d\lambda^2} \tag{1}$$

where $D(\lambda)$ is the material dispersion parameter (usually given in ps/nm·km), $\Delta\lambda$ is the spectral width of the light source, and L is the length of the fiber. Fig. 3(a) depicts a typical material dispersion curve as a function of the operating wavelength. for a PF GIPOF as well as a silica-based MMF with a SiO$_2$ core doped with 6.3mol-% GeO$_2$ and a SiO$_2$ cladding. It is clearly seen the better performance in terms of material dispersion of the PF GIPOF compared to the silica-based counterpart, especially in the range up to 1100nm.

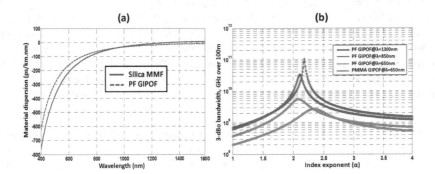

Figure 3. (a) Typical material dispersion of the central core region for a silica-based MMF (blue solid line) and PF GIPOF (red dashed line). (b) Relation between the refractive index profile and bandwidth of 100m-long PF GIPOF. PMMA-GIPOF at 650nm is plotted for comparison.

On the other hand, modal dispersion is caused by the fact that the different modes (light paths) within the fiber carry components of the signals at different velocities, which ultimate results in pulse overlap and a garbled communications signal. Lower order modes propagate mainly along the waveguide axis, while the higher-order modes follow a more zigzag path, which is longer. If a short light pulse is excited at the input of the fiber, the lowest order modes arrive first at the end of the fiber and the higher order modes arrive later. The output pulse will thus be built up of all modes, with different arrival times, so the pulse is broadened.

To overcome and compensate for modal dispersion, the refractive index of the fiber core (or, alternatively, graded index exponent of the fiber core) is graded parabola-like from a high index at the fiber core center to a low index in the outer core region, i.e. by forming a graded-index (GI) fiber core profile. In such fibers, light travelling in a low refractive-index structure has a higher speed than light travelling in a high index structure and the higher order modes bend gradually towards the fiber axis in a shorter period of time because the refractive index is lower at regions away from the fiber core. The objective of the GI profile is to equalise the propagation times of the various propagating modes. Therefore, the time difference between the lower order modes and the higher order modes is smaller, and so the broadening of the pulse leaving the fiber is reduced and, consequently, the transmission bandwidth can be increased over the same transmission length. For negligible modal dispersion the ideal refractive index profile is around 2. This refractive index profile formed in the core region of multimode optical fibers plays a great role determining its bandwidth, because modal dispersion is generally dominant in the multimode fiber although an optimum refractive index profile can produce the minimum modal dispersion, i.e. larger bandwidth being almost independent of the launching conditions [18]. Fig. 3(b) shows the calculated bandwidth of a PF-based GIPOF operating at different wavelengths, in which it is assumed that the source spectral width is 1nm, with regards to the refractive index profile, α. The data of the bandwidth of a PMMA-based GIPOF at 650nm is also shown for comparison showing a maximum limited to approximately 1.8GHz for 100m by the large material dispersion. On the other hand, the smaller material dispersion of the PF polymer-based GIPOF permits a maximum bandwidth of 4GHz even at 650nm. Furthermore, when the signal wavelength is 1300nm, theoretical maximum bandwidth achieves 92GHz for 100m. The difference of the optimum index exponent value between 650nm and 1300nm wavelengths is caused by the inherent polarization properties of material itself. It should be mentioned that a uniform excitation has been assumed and no differential mode attenuation (DMA) and mode coupling (MC) effects have been considered. These effects will be briefly described later on.

To summarize, the different types of dispersion that appear in a MMF and their relation to the fiber bandwidth are analyzed in Fig. 4. This figure reports the PF GIPOF chromatic and modal dispersion and the total bandwidth of a 100m-long link as a function of the refractive index profile, at a wavelength of 1300nm. Fig. 4(b) depicts the corresponding 3-dBo (3-dB optical bandwidth) baseband bandwidth, related to Fig. 4(a). These plots are based on the same analysis of Fig. 3, which assumed a uniform excitation and neglected both the DMA and mode coupling effects. From these figures, the chromatic bandwidth is seen to show little dependence on α, which means that the material dispersion is the dominant contribution (with regards to the profile dispersion) in the transmission window considered. On the other hand, the modal bandwidth shows a highly peaked resonance with α. This is the well known characteristic feature of the grading. With the present choice of parameters values, that maximum bandwidth (i.e. minimum dispersion) approximately occurs at 2.18 at 1300nm, as shown in Fig. 4(a). Furthermore, the presence of crossover points (namely α_1 and α_2) shows that the total bandwidth may be limited either by the modal dispersion or the chromatic dispersion depending on the value of the refractive index profile. Focusing on

Fig. 4(a), the chromatic dispersion will essentially limit the total bandwidth for $\alpha_1 < \alpha < \alpha_2$, whilst for $\alpha < \alpha_1$ or $\alpha > \alpha_2$ the modal dispersion will cause the main limitation. In other words, when the index exponent is around the optimum value (α-resonance), the modal dispersion effect on the possible 3-dB bandwidth (and so on the bit rate) is minimized and the chromatic dispersion dominates this performance. On the other hand, when the index exponent is deviated from the optimum, the modal dispersion increases becoming the main source of bandwidth limitation.

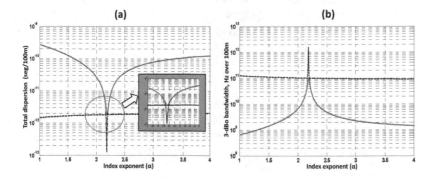

Figure 4. (a) Dispersion effects versus refractive index profile for a 100m-long PF GIPOF, assuming equal power in all modes and a 1300nm light source with 1nm of spectral linewidth. Inset: zoom near the optimum profile region. (b) Corresponding 3-dBo bandwidth. (—) Total dispersion ; (- -) Modal dispersion ; (---) Chromatic dispersion.

It is also noteworthy that, since the PF polymer has low material and profile dispersions and the wavelength dependence of the optimum profile is decreased, a high bandwidth performance can be maintained over a wide wavelength range, compared to multimode silica or PMMA-based GIPOF fibers.

2.2.1. Dispersion modelling approach

The propagation characteristics of optical fibers are generally described by the wave equation which results directly from Maxwell's equations and characterizes the wave propagation in a fiber as a dielectric wave guide in the form of a differential equation. In order to solve the equation, the field distributions of all modes and the attendant propagation constants, which results from the use of the boundary conditions, have to be determined.

The wave equation is basically a vector differential equation which can, however, under the condition of weak wave guidance be transformed into a scalar wave equation in which the polarization of the wave plays no role whatsoever [19]. The prerequisite for the weak wave guiding is that the refractive indices between the core and cladding hardly differ, being fulfilled quite well in silica fibers when the difference in refractive index between the core and cladding region is below 1%. Calculations based on the scalar wave equation only show very small inaccuracies with regards to the group delay. Then, the equations which describe the electric and magnetic fields are decoupled so that you can write a scalar wave equation.

The models based on the solution of the wave equation in the form of a mode solver differ fundamentally only in regard to the solution method and whether or not you are proceeding from a more computer-intensive vector wave equation or the more usual scalar wave equation. In the technical literature solutions for the vector wave equation with the aid of finite element method (FEM) [20], with finite differences (Finite Difference Time Domain Method - FDTD) [21] and the beam propagation method (BPM) [22] are well known. These are generally used for very small, mostly singlemode waveguides in which polarization characteristics play a role. Multimode fibers (including polymer fibers) are quite large and the polarization of light counts for only a few centimeters. That is why analytical estimations of the scalar wave equation, the so-called WKB (Wentzel-Kramers-Brillouin, from whom the name derives) Method and Ray Tracing [23], are primarily used for the modeling of multimode fibers. In the latter, the propagating light through an optical system can be seen as the propagation of individual light rays following a slightly different path; these paths can be calculated using standard geometrical optics.

Focusing on the WKB method, the latter primarily makes available expressions, that can be calculated efficiently, for describing the propagations constants and group delays of the propagating modes within the fiber. In this method, whereas the field distributions in step index profile fibers can be determined analytically, the refractive index distribution over the radius of a graded index fiber can generally be described with a power-law, as Eq. 2 states. Fibers with power-law profiles possess the characteristic that the modes can be put in mode groups which have the same propagation constant and also similar mode delay (at least for exponents close to $\alpha = 2$). The propagation times of the modes are only then dependent on the propagation constant and then the group delay can be determined with the aid of the WKB Method by differentiating the propagation constant from the angular frequency [24].

$$n(r,\lambda) = \begin{cases} n_1(\lambda)\left[1 - 2\Delta(\lambda)\left(\dfrac{r}{a}\right)^{\alpha}\right]^{1/2} & \text{for } 0 \le r \le a \\ n_1(\lambda)\left[1 - 2\Delta(\lambda)\right]^{1/2} & \text{for } r \ge a \end{cases} \quad \text{with } \Delta(\lambda) = \dfrac{n_1^{\,2}(\lambda) - n_2^{\,2}(\lambda)}{2n_1^{\,2}(\lambda)} \qquad (2)$$

where r is the offset distance from the core center, a is the fiber core radius (i.e. the radius at which the index $n(r, \lambda)$ reaches the cladding value $n_2(\lambda) = n_1(\lambda)[1 - 2\Delta(\lambda)]^{1/2}$), $n_1(\lambda)$ is the refractive index in the fiber core center, λ is the free space wavelength of the fiber excitation light, α is the refractive index exponent and $\Delta(\lambda)$ is the relative refractive index difference between the core and the cladding. It is usually assumed that the core and cladding refractive index materials follow a three-term Sellmeier function of wavelength [25] given by:

$$n_i(\lambda) = \left(1 + \sum_{k=1}^{3} \frac{A_{i,k}\lambda^2}{\lambda^2 - \lambda_{i,k}^2}\right)^{1/2} \quad \text{with i=1 (core), 2 (cladding)} \qquad (3)$$

where $A_{i,k}$ and $\lambda_{i,k}$ are the oscillator strength and the oscillator wavelength, respectively (both parameters are often gathered under the term of Sellmeier constants).

On the other hand, from the WKB analysis, the modal propagation constants can be approximately derived as following [26], in which each guided mode has its own propagation constant and therefore propagates at its own particular velocity:

$$\beta_m = \beta(m, \lambda) = n_1(\lambda)k \left[1 - 2\Delta(\lambda) \left(\frac{m}{M(\alpha, \lambda)} \right)^{\frac{2\alpha}{\alpha+2}} \right]^{1/2} \tag{4}$$

where m stands for the principal mode number [27] and $k = 2\pi / \lambda$ is the free space wavenumber. This so-called principal mode number (mode group number or mode number) can be defined as $m = 2\mu + v + 1$ in which the parameters μ and v are referred to as radial and azimuthal mode number, respectively. Physically, μ and v represent the maximum intensities that may appear in the radial and azimuthal direction in the field intensities of a given mode. For a deeper analysis works reported in [28, 29] are recommended. On the other hand, $M(\alpha, \lambda)$ is the total number of mode groups that can be potentially guided in the fiber, given by [26]:

$$M(\alpha, \lambda) = 2\pi a \frac{n_1(\lambda)}{\lambda} \left[\frac{\alpha \cdot \Delta(\lambda)}{\alpha + 2} \right]^{1/2} \tag{5}$$

As a consequence of Eq. 4, the delay time $\tau(m, \lambda)$ of a mode depends only on its principal mode number. It should be mentioned that the differences in modal delay are those that determine the modal dispersion. The delay time of the guided modes (or modal delay per unit length) can be derived from Eq. 4 using the definition:

$$\tau(m, \lambda) = -\frac{\lambda^2}{2\pi c} \frac{d\beta(m, \lambda)}{d\lambda} \tag{6}$$

where c is the speed of light in vacuum, deriving in:

$$\tau_m = \tau(m, \lambda) = \frac{N_1(\lambda)}{c} \left[1 - \frac{\Delta(\lambda)(4 + \varepsilon(\lambda))}{\alpha + 2} \left(\frac{m}{M} \right)^{\frac{2\alpha}{\alpha+2}} \right] \left[1 - 2\Delta(\lambda) \left(\frac{m}{M} \right)^{\frac{2\alpha}{\alpha+2}} \right]^{-1/2} \tag{7}$$

where $\varepsilon(\lambda)$ is the profile dispersion parameter given by [30]:

$$\varepsilon(\lambda) = -\frac{2n_1(\lambda)}{N_1(\lambda)}\frac{\lambda\frac{d\Delta(\lambda)}{d\lambda}}{\Delta(\lambda)} \tag{8}$$

and $N_1(\lambda)$ is the material group index defined by:

$$N_1(\lambda) = n_1(\lambda) - \lambda\frac{dn_1(\lambda)}{d\lambda} \tag{9}$$

2.3. Differential mode attenuation

The distribution of the power among the different modes propagating through the fiber will also be affected by the Differential Mode Attenuation (DMA), also called mode-dependent attenuation, which causes the attenuation coefficient to vary from mode to mode in a different manner. It originates from conventional loss mechanisms that are present in usual optical fibers such as absorption, Rayleigh scattering [31] or losses on reflection at the core-cladding interface [32]. The following functional expression or empirical formula for the DMA is proposed, in which the DMA increases when incresing the mode order [33]:

$$\alpha_m = \alpha_m(m,\lambda) = \alpha_o(\lambda) + \alpha_o(\lambda)I_\rho\left[\eta\left(\frac{m-1}{M}\right)^{\frac{2\alpha}{\alpha+2}}\right] \tag{10}$$

where $\alpha_o(\lambda)$ is the attenuation of low-order modes (i.e intrinsic fiber attenuation), I_ρ is the ϱ-th order modified Bessel function of the first kind and η is a weighting constant. This empirical formula is set up by noticing that most measured DMA data displayed in the literature for long wavelengths conform to the shape of modified Bessel functions [31, 34, 35]. It is also worth mentioning that, during propagation, modes with fastest power loss may be stripped off or attenuated so strongly that they no longer significantly contribute to the dispersion. In other words, the DMA is a filtering effect, which may yield a certain bandwidth enhancement depending on the launching conditions and the transmission length. From Fig. 5 it can be seen that low-order mode groups show similar attenuation (intrinsic fiber attenuation) whereas for high-order mode groups attenuation increases rapidly.

2.4. Mode coupling

Mode coupling is rather a statistical process in which modes exchange power with each other. Due to the mode coupling, the optical energy of the low-order modes would be coupled to higher-order modes, even if only the low-order modes would have launched selectively. This effect generally occurs through irregularities in the fiber, whether they are roughness of the core-cladding interface or impurities in the core material leading, for instance, to refrac-

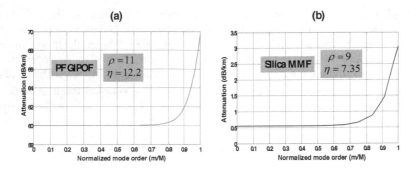

Figure 5. (a) Differential mode attenuation (DMA) as a function of the normalized mode order m/M for a PF GIPOF with a=250µm, α=2, and λ=1300nm. An intrinsic attenuation of 60dB/km@1300nm has been considered. (b) Differential mode attenuation (DMA) as a function of the normalized mode order m/M for a silica MMF with a=31.25µm, α=2, and λ=1300nm. An intrinsic attenuation of 0.55dB/km@1300nm has been considered.

tive index fluctuations. This effect can therefore only be described with statistical means. In addition, it is agreed that silica-based MMFs exhibit far less mode coupling compared to POF fibers [36]. This is attributed to the difference in the material properties.

The main effects for generating mode coupling are Rayleigh and Mie scattering which differ in the size of the scattering centers. Rayleigh scattering arises through the molecular structure of matter which is why no material can have perfectly homogenous properties. Its optical density fluctuates around a mean value which represents the refractive index of the material. These fluctuations are very small and have typical sizes in the range of molecules (<µm). Rayleigh scattering depends on the wavelength and decreases with greater wavelengths as of the fourth power ($\sim \lambda^{-4}$). In constrast, Mie scattering comes from the fluctuations of the refractive index which has greater typical lengths that mostly come about because of impurities in the material such as air bubbles or specks of dust which are large compared with the wavelength of light. The ensuing scattering has more of an effect on the direction of propagation of the light and is independent of the wavelength. Thinking of these aspects mode coupling reveals itself as a complex process which plays a great role in polymer fibers.

There are some approaches for the modeling of mode coupling which cannot be applied equally well in all propagation models [37, 38] while some descriptions present themselves rather in mode models [39]. Moreover, the coupling coefficients which describe the coupling between modes can either be described by analytical attempts which are based on observations of mode overlapping [40, 41]. However, it is demonstrated that in real fibers only very few modes effectively interact with each other and, moreover, neighboring or adjacent modes (those with similar propagation constants, modes m and $m\pm1$, respectively) primarily show strong mode coupling [42, 43]. As a matter of fact, larger core refractive index and higher fiber numerical aperture (NA) values are expected to decrease the mode coupling in GIPOFs. In addition, larger mode coupling effects are observed in SIPOFs compared to that GIPOFs counterparts.

Mode coupling alters the achievable bandwidth of a multimode fiber. According to the laws of statistics, the differential delay (or more precisely, the standard deviation) between the different propagating modes does not increase in a linear relationship to the length but approximately only proportional to the square root of the length. The best known approach for approximately determining the coupling length of the fiber is the description with the aid of a length-dependent bandwidth, in the way $BW \propto L^{\gamma}$. Here the coupling length is the point in which the linear decrease ($\gamma \approx -1$) in the bandwidth turns to a root dependency ($\gamma \approx -0.5$) under mode coupling. From this point, a state of equilibrium arises through mode coupling effects. Typical values of coupling length in silica-based GI-MMFs are in the order of units of kilometers [44] whereas in the case of PF GIPOFs usually range from 50m up to 150m.

3. Multimode optical fiber capabilities

Emerging themes in next-generation access (NGA) research include convergence technologies, in which wireline-wireless convergence is addressed by Radio-over-Fiber (RoF) technologies. Photonics will transport gigabit data across the access network, but the final link to the end-user (measured in distances of metres, rather than km's) could well be wireless, with portable/mobile devices converging with photonics. RoF technologies can address the predicted multi-Gbps data wave, whilst conforming to reduced carbon footprints (i.e. green telecoms). NGA networks will provide a common resource, with passive optical networks (PONs) supplying bandwidth to buildings, and offering optical backhaul for such systems.

Parameter	Remarks
Transmission distances	Typ. <10km, max. 20km, e.g. for alternative topologies
Peak data rate	100Mbps (private customers)
	Nx1 Gbps up to 10Gbps (business)
Temperature range	Controlled: +10ºC to +50ºC
	Uncontrolled operation in buildings: -5ºC to +85ºC
	Uncontrolled operation in the field: -33ºC to +85ºC
Long lifetime	
Humidity and vibrations (shock) have to be considered at non-weather protected locations	
No optical amplifiers in the field	
No optical dispersion compensation	

Table 1. Access network requirements.

It has been indicated by several roadmaps that the peak link data rate should be at least 100Mbps (symmetrical) for private customers and 1 to 10Gbps for business applications. Inherent access network requirements are highlighted in Table 1. These hundreds of megabits per second per user are reasonably reachable in the coming future and the Fiber To The

Home (FTTH, or some intermediate version such as FTT-curb) network constitutes a fiber access network, connecting a large number of end users to a central point, commonly known as an access node. Each access node will contain the required active transmission equipment used to provide the applications and services over optical fiber to the subscriber.

On the one hand, Ethernet is the most widespread wired LAN technology, including in-home networks, and the development of Ethernet standards goes hand in hand with the adoption and development of improved MMF channels [45]. And Ethernet standards for 1Gbps and 10Gbps designed for multimode and singlemode fibers are now in use. Table 2 shows the minimum performance specified by IEEE 802.3 standard for the various interfaces. For example, 10-Gigabit Ethernet (GbE) standard operating at 10.3125Gbps@1300nm supports a range of transmission lengths of 300m over multimode silica fiber and 10km over singlemode silica fiber. Actually OM4 fiber type is under consideration although is not yet within a published standard. OM4 fiber type defines a 50μm core diameter MMF with a minimum modal bandwidth (under OverFilled Launching condition, OFL) of 3500MHz· km@850nm and 500MHz·km@1300nm, respectively. Nevertheless, data rate transmission research achievements are not at par as those covered by the standard and report even greater values. Some significant works are reported in [46-48]. Different techniques or even a combination of some of them were applied to achieved these transmission records. Some of them will be briefly discussed in next section.

Fiber type	10GBaseSR 850nm Modal Bandwidth / Operating Range (MHz·km)/(Meters)	10GBaseLR(/ER) 850nm Modal Bandwidth / Operating Range (MHz·km)/(Km)	10GBaseLRM 1300nm Modal Bandwidth /Operating Range (MHz·km)/ (Meters)
62.5μm*	160/26	n.a.	n.a.
62.5μm (OM-1)**	200/33	n.a.	500/300
50μm	400/66	n.a.	400/240
50μm (OM-2)	500/82	n.a.	500/300
50μm (OM-3)	2000/300.	n.a.	500/300
SMF	n.a.***	10 (/40)	n.a/10000.

*TIA (Telecommunications Industry Association), Document 492AAAA compliance. Commonly referred to as 'FDDI-grade' fiber.

** ISO (International Standards Organization), Document 11801 compliance.

*** n.a.: not available.

Table 2. 10-Gigabit Ethernet transmission over fiber standards (IEEE 802.3aq). Approved in 2006.

Figure 6 provides a brief description of the current 10GbE and the possible future Ethernet standards over copper and fiber links [6]. The trend of extending the reach and data rate of the links is obvious in the previous standards and the 10GbE standards shown in the figure. Although the twisted pair of copper wires is a relatively low-cost and low-power solution compared to the MMF solutions, the motivation for the transition from the copper-based links to the MMF links is their much higher available bandwidth. However, the need for even higher performance MMF solutions is apparent, and much more is to be expected, for example, with new ultra-HDTV format such as 4K (4000 horizontal pixels, with an expected increase in the required bandwidth of a factor of approximately 16).

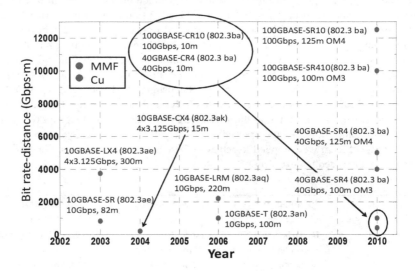

Figure 6. Gigabit Ethernet (10GbE) standards over MMF and copper links [44].

On the other hand, another important point in access networks communications is within the field of the wireless signal transmission (for both mobile and data communication), namely Wireless Local Area Networks (WLANs). Wireless technologies are developing fast but there is a need to link base stations/servers to the antenna by using fixed links together with the future exploitation of capacities well beyond present day standards (IEEE802.11a/b/g), which offer up to 54Mbps and operate at 2.4GHz and 5GHz, as well as 3G mobile networks such as IMT2000/UMTS[4], which offer up to 2Mbps and operate around 2GHz. Moreover, IEEE802.16, otherwise known as WiMAX, is another recent standard aiming to bridge the last mile through mobile and fixed wireless access to the end user at frequencies between 2-11GHz. In addition, WiMAX also aims to provide Fixed Wireless Access at bit-rate in the excess of 100Mbps and at higher frequencies between 10-66GHz. All these services use signals at the radio-frequency (RF) level that are analogue in nature, at least in the sense that they cannot be carried directly by digital baseband modulation. Optical cabling solutions can also offer the possibility for semi-transparent transport of these signals

---

4 IMT2000: International Mobile Telecommunications-2000 ; UMTS:Universal Mobile Telecommunications System.

by using Radio-over-Fiber (RoF) technology. This RoF technology has been proposed as a solution for reducing overall system complexity by transferring complicated RF modem and signal processing functions from radio access points (RAPs) to a centralised control station (CS), thereby reducing system-wide installation and maintenance costs. Furthermore, although RoF in combination with multimode fibers can be deployed within homes and office buildings for baseband digital data transmission within the Ultra Wide Band (UWB), in general low carrier frequencies offer low bandwidth and the 6GHz UWB unlicensed low band is not available worldwide due to coexistence concerns [49]. These include radio and TV broadcasts, and systems for (vital) communication services such as airports, police and fire, amateur radio users and many others. In contrast, the 60 GHz-band, within millimetre wave, offers much greater opportunities as the resulting high radio propagation losses lead to numerous pico-cell sites and thus to numerous radio access points due to the limited cell coverage. These pico-cells are a natural way to increase capacity (i.e. to accommodate more users) and to enable better frequency spectrum utilisation. Therefore, for broadband wireless communication systems to offer the needed high capacity, it appears inevitable to increase the carrier frequencies even to the range of millimetre-wave and to reduce cell sizes [50]. Considering in-house wireless access networks, coaxial cable is very lossy at such frequencies and the bulk of the installed base of in-building fiber is silica-based MMF. Meanwhile PF GIPOF is also emerging as an attractive alternative, due to the aforementioned low cost potential and easier handling required in in-building networks. It is also mandatory to overcome the modal bandwidth limitation in multimode fibers to deliver modulated high frequency carriers to remote access points.

Following on this, it should be mentioned that PF GIPOFs have been demonstrated capable for transmission of tens of Gbps over distances of hundreds of meters. Some examples are reported in [51-53] in which more than 40Gbps over 100m of PF GIPOF are reported. An overview of some significant works over the years regarding GIPOF transmission can be seen in [54]. This is in contrast with all commercially available step-index POFs (SIPOFs) in which the bandwidth of transmission is limited to about 5MHz·km [6] due to modal dispersion. Therefore, even in the short-range communication scenario, the SIPOF is not able to cover the data rate of more than 100Mbps that would be necessary in many standards of the telecommunication area. Therefore, the SIPOF is mainly aimed at very short-range data-range transmission (less than 50m), image guiding and illumination.

3.1. Multimode optical fiber expanded capabilities

Although multimode fibers, both silica-based and polymer-based counterparts, are the best candidate for the convergence and achievement of a full service access network context, it has been previously addressed their main disadvantage concerning the limited bandwidth performance, limited by modal dispersion. For instance, for standard 62.5/125μm silica-based MMFs, the minimum bandwidths are only specified to be 200MHz·km and 500MHz·km (up to 800MHz·km) in the 850nm and 1300nm transmission windows, respectively, under OverFilled Launch (OFL) condition[5]. Even though these specifications do satisfy the information rate of many classical short-range links, it is clear that a 2km-long campus backbone cannot be realized for operation at the speed of Gigabit Ethernet. This limited bandwidth hampers the desired integration of multiple broadband services into a common multimode fiber access or in-building/home network. Overcoming the bandwidth limitation of such fibers requires the development of techniques oriented to extend the capabilities of multimode fiber networks to attend the consumer's demand for multimedia services.

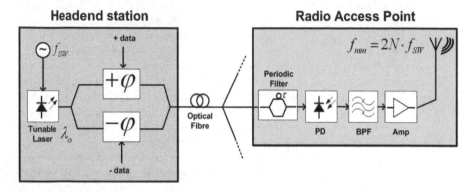

Figure 7. Feeding microwave data signals over a multimode network by OFM technique.

Novel techniques to expand the MMF capabilities and surmount this bandwidth bottleneck are continuously reported demonstrating that the frequency response of MMF does not diminish monotonically to zero after the baseband bandwidth, but tends to have repeated passbands beyond that [55]. In recent times, these high-order passbands and flat regions have been used in research to transmit independent streams of data (digital or analogue) complementary to the baseband bandwidth in order to exceed the aggregated transmission capacity of MMF [56] as well as to transport microwave and mm-wave radio carriers, commonly employed for creating high-capacity picocell wireless networks in RoF systems, as in [57]. Related to this latter technique, the Optical Frequency Multiplying (OFM) is a method by which a low-frequency RF signal is up-converted to a much higher microwave frequency through optical signal processing [58]. At the headend station, a wavelength-tunable optical source is used, of which the wavelength is periodically swept over a wavelength range with a sweep f_{SW} while keeeping its output power constant. The data is then impressed on this wavelength-swept optical signal, see Fig. 7. After having passed through the optical fiber link, the signal impinges on a periodic optical multi-passband filter (e.g. optical comb or Fabry-Perot filter). In sweeping across N transmission peaks of this filter (back and forth during one wavelength sweep cycle), light intensity burts arrive on the photodiode with a frequency $2 \cdot N \cdot f_{SW}$. Thus, the output signal of the photodiode contains a microwave frequency component at the above frequency and higher harmonics of which the strength depends on the bandpass characteristics of the periodic filter. Then, in order to select the desired har-

---

5 ISO/IEC (International Standards Organization/International Electrotechnical Commission) 11801-"Generic cabling for customer premises".

monic, a bandapss filtering plus some amplification could be implemented. Note that only the optical sweep frequency is limited by the bandwidth of the optical fiber link, and that microwave carrier frequency can exceed this bandwidth by far due to the optical frequency multiplication mechanism. Extremely pure generated microwave signals have been demonstrated, notwithstanding a moderate laser spectral linewidth, due to the inherent phase noise cancellation in the OFM technique [59, 60].

On the other hand, subcarrier multiplexing (SCM) is a mature, simple, and cost effective approach for exploiting optical fiber bandwidth in analogue optical communication systems in general and RoF systems in particular. This technique was firstly addressed at the end of the 1990's in [61], which also takes advantage of the relative flat passband channels existing in the multimode fiber frequency response. Basically, in SCM, the RF signal (the subcarrier) is used to modulate an optical carrier at the transmitter's side. As a result, there is an optical spctrum consisting of the original optical carrier f_0 plus two side-tones located at $f_0 \pm f_{SC}$, where f_{SC} is the subcarrier frequency. If the subcarrier itlsef is modulated with data (either analogue or digital), then sidebands centered on $f_0 \pm f_{SC}$ are produced. Finally, to multiplex multiple channels on to one optical carrier, multiple subcarriers are first combined and then used to modulate the optical carrier [62]. At the receiver's side the sucarriers are recovered through direct detection. One of the main advantages of SCM is that it supports broadband mixed mode data traffic with independent modulation format. Moreover, one subcarrier may carry digital data, while another may be modulated with an analogue signal, such as telephone traffic. However, the frequency ranges suitable for passband transmission vary from fiber to fiber as well as with the fiber length, the launching conditions or if the fiber is subjected to mechanical stress. Nevertheless, to overcome this limitation, an adaptive channel/allocation system would be necessary. Another drawback is that being SCM an analogue communication technique, it becomes more sensitive to noise effects and distortions due to non-linearities in the communications system.

It is worth noting that some other methods try to electrically improve this bandwidth performance using, for example, equalization techniques [63, 64]. In addition to, it is well known than an m-ary digital modulation scheme with m>2 (multi-level coding) can enhance transmission capacity by overcoming the bandwidth limitations of a transmitter or a transmission medium and, therefore, multilevel modulation schemes that are used in radio-frequency communications have also been demonstrated in fiber-optic links [65]. Other attempts to overcome the bandwidth limit includes selective excitation of a limited number of modes, so-called Restricted Mode Launching (RML), in different ways: offset launch [66], conventional center launch [67] or even by means of a twin-spot technique [68]. Since the propagating modes are fewer under RML launch conditions, the difference in propagating times between the fastest and slowest modes is smaller, thus decreasing modal dispersion and increasing the corresponding bandwidth. In a similar way, Mode Group Diversity Multiplexing (MGDM) [69, 70] can be applied, in which the bandwidth increase is achieved by injecting a small light spot radially offset from the fiber core center thus limiting the number of modes excited within the fiber and, therefore, performing different simultaneous data transmission channels depending on the group of modes propagating. On the other hand, from the multimode fiber frequency response, the effect of having a wideband frequency-selective channel for data transmission can be overcome by using orthogonal frequency-division multiplexing (OFDM). In OFDM, the high-data-rate signal is error-correction encoded and then divided into many low-data-rate signals. By doing this, the wideband frequency-selective channel is separated into a series of many narrowband frequency-nonselective channels. OFDM technique has been applied to fiber-optic transmission [71] and shown to offer some protection against the frequency selectivity of a dispersive multimode fiber. Mode filtering techniques, either at the fiber input [72] or its output [73] have also been applied.

As cost is a key issue in local and residential networks, the use of Wavelength Division Multiplexing Passive Optical Network (WDM-PON) architectures for distribution of RoF signals has gained importance recently as WDM enables the efficient exploitation of the fiber network's bandwidth. This architecture acts as the starting point from the access node to the subscribing homes and buildings, constituting the all-optical fiber path. WDM-PON promises to combine both sharing feeder fibers while still providing dedicated point-to-point connectivity [74]. A basic scheme of the WDM-PON architecture can be seen in Fig. 1(b). In this case, optical microwave/mm-wave signals from multiple sources, which can be located in a Central Office (CO) or Optical Line Terminal (OLT) can be multiplexed and the composite signal is transported through an optical fiber and, finally, demultiplexed to address each Optical Network Terminal (ONT) or Remote Access Point (RAP), the latter for wireless applications. However, a challenging issue concerns the applications of these signals as the optical spectral width of a single mm-wave source may approach or exceed the WDM channel spacing.

Finally, it is worth mentioning that there is not the desire of making a competition between optical and wireless solutions, since wireless is and will always be present inside the building or home. In contrast, research and development are focusing on the coexistence of both technologies.

4. Theoretical approach of multimode optical fibers

4.1. Introduction

The restricted bandwidth of the multimode fiber has been one of the main causes that makes the specification and designing of the physical media dependent layer very difficult. Moreover, the potentials of MMFs to support broadband RF, microwave and millimetre wave transmission over short, intermediate and long distances to meet user requirements for higher data rates and to support emerging multimedia applications are yet to be fully known. To enable the design and utilization of MMFs with such enhanced speeds, the development of an accurate frequency response model to describe the signal propagation through multimode fibers is of prime importance. Through this multimode fiber modelling more likely performance limits can be established, thereby preventing eventual overdesign of systems and the resulting additional cost.

Since the mid-1970's, much work has been directed to the investigation of MMFs and their ability for high speed transmission. Different factors have clearly been identified to influence the information-carrying capacity, namely the material dispersion (in combination with the spectrum of the exciting source) [26], the launching conditions [66] as well as the mode-dependent characteristics, i.e. delay [26], attenuation [75] and coupling coefficient [27]. Unfortunately, the achievements, so far accomplished, are not quite complete to enable precise frequency response and bandwidth prediction if an arbitrary operating condition is to be considered.

The most popular technique reported so far for the analysis of signal propagation through MMF fibers is that based on the coupled power-flow equations developed by Gloge [76] in the early 70's and later improved by Olshansky [27] and Marcuse [28], to account for the propagation and time spreading of digital pulses through MMFs. Most of the published models and subsequent work on the modelling of MMFs [29, 77-79] are based on this method in which the MMF power transfer function is solved by means of a numerical procedure like the Crank-Nicholson method, for instance [29]. However, other methods rely on solving the system of coupled equations adopting the matrix formalism [80].

The power-flow equations are adequate for the description of digital pulse propagation through MMFs but present several limitations either when considering the propagation of analogue signals or when a detailed knowledge of the baseband and RF transfer function is required since in these situations the effect of the signal phase is important. To overcome these limitations it is necessary to employ a method relying on the propagation of electric field signals rather than optical power signals. Unfortunately, there are very few of such descriptions available in the literature with the exception of the works reported in [54, 81, 82].

From literature, it is demonstrated that the frequency characteristics of multimode fibers should show significant high-frequency components, i.e. higher-frequency transmission lobes, resonances or passbands are expected in the fiber frequency response. And these higher-frequency transmission lobes would allow to transport information signals by modulating them on specific carrier frequencies, as an independent transmission channel each. These modulated carriers can be positioned in such a way that they will optimally fit into the higher-frequency transmission lobes of the multimode fiber link thus increasing the aggregated transmission capacity over MMFs. Furthermore, it has been stated that the contrast ratio between resonances reveals a dramatically reduction as the frequency increases thus providing potential for broadband transmission at even higher frequencies than those determined by the transmission lobes.

The position of these higher-frequency lobes depends on the fiber link length, and on the exact fiber characteristics, which may vary due to external circumstances such as induced stress by bending or environmental temperature variations. Any system that would take advantage of such high-frequency transmission lobes would have to adapt to those variations, e.g. monitoring the fiber link frequency response by injecting some weak pilot tones, and allocating the subcarriers accordingly would be a feasible solution. Anyway, this in turn is contingent on the availability of accurate models to describe the microwave radio signals propagation over multimode fibers. With such a predictive tool, notwithstanding its restricted bandwidth, a single multimode fiber network that may carry a multitude of broadband services using the higher-order transmission lobes would become more feasible. Thus, easy-to-install multimode fiber networks for access and in-building/home can be realised in which wirebound and wireless services were efficiently integrated.

4.2. Mathematical framework

In this section a closed-form analytic expression to compute the baseband and RF transfer function of a MMF link based on the electric field propagation method is briefly presented. By obtaining an accurate model it is possible to evaluate the conditions upon which broadband transmission is possible in RF regions far from baseband. For a deeper comprehension works reported in [81, 82] are recommended.

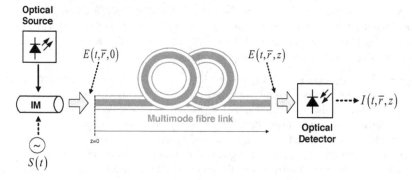

Figure 8. Scheme of a generic Multimode Optical Fiber link. IM: Intensity optical Modulator.

Fig. 8 shows a generic optical transmission system scheme which employs a multimode optical fiber as a transmission medium. $E(t, \bar{r}, z)$ represents the electric field at a point located at a distance z from the fiber origin and at a point $\bar{r}$ of its cross section. $E(t, \bar{r}, 0)$ represents the electric field at the fiber origin and at a point $\bar{r}$ of its cross section and $S(t)$ is the modulation signal composed of a RF tone with modulation index m_o.

Thus the optical intensity at a point z, $I(t, \bar{r}, z)$, depends directly on the electric field $E(t, \bar{r}, z)$ at a point located at a distance z from the fiber origin and at a point $\bar{r}$ of its cross section. Both the electric field and the optical intensity can be expressed, using the electric field propagation model and referred to the system described in Fig. 8, as [82]:

$$E(t, \bar{r}, z) = \sum_{\nu=1}^{N} \sum_{\mu=1}^{N} [h_{\mu\nu}(t) * E_\nu(t, 0)] e_\nu(\bar{r}) \tag{11}$$

$$I(t, \bar{r}, z) \propto \left\langle \left| E(t, \bar{r}, z) \right|^2 \right\rangle = \sum_{\mu=1}^{N} \sum_{\nu=1}^{N} \sum_{\mu'=1}^{N} \sum_{\nu'=1}^{N} e_\nu^*(\bar{r}) e_{\nu'}(\bar{r}) \cdot$$

$$\cdot \int_{-\infty}^{\infty} \int_{-\infty}^{\infty} \left\langle h_{\mu\nu}^*(t - t') h_{\mu'\nu'}(t - t'') \right\rangle \left\langle E_\mu^*(t', 0) E_{\mu'}(t'', 0) \right\rangle dt' dt'' \tag{12}$$

where N is the number of guided modes, $h_{\mu\nu}(t)$ is the impulse response at z caused by mode ν at the fiber origin over mode μ at z and $e_\nu(r)$ is the modal spatial profile of mode ν. It has been assumed that non linear effects are negligible.

Let $S(t)$ be the modulation signal composed of a RF tone with modulation index m_o assuming a linear modulation scheme (valid for direct and external modulation), which incorporates the source chirp α_C, and approximated by three terms of its Fourier series, following:

$$\sqrt{S(t)} = \sqrt{S_o}\left\{1 + \frac{m_o}{8}(1 + j\alpha_C)e^{j\Omega t} + \frac{m_o}{8}(1 + j\alpha_C)e^{-j\Omega t}\right\} \tag{13}$$

where S_o is proportional to the average optical power and Ω represents the frequency of the RF modulating signal. It has also been assumed an optical source which has a finite linewidth spectrum (temporal coherence) defined by a Gaussian time domain autocorrelation function.

Assuming a stationary temporal coherence of the source and assuming that the detector collects the light impinging on the detector area A_r, and produces an electrical current proportional to the optical power given by:

$$P(t) = \int_{A_r} I(t, \bar{r}, z)d\bar{r} = \int_{-\infty}^{\infty}\int_{-\infty}^{\infty}\sqrt{S^*(t')S(t'')} \cdot Q(t - t', t - t'')dt'dt'' \tag{14}$$

being

$$Q(t', t'') = R(t', t'')Q_0(t', t'') \text{ and } Q_0(t', t'') = \sum_{\mu=1}^{N}\sum_{\nu=1}^{N}\sum_{\mu'=1}^{N}\sum_{\nu'=1}^{N}C_{\mu\mu'}\chi_{\nu\nu'}\left\langle h^*_{\mu\nu}(t')h_{\mu'\nu'}(t'')\right\rangle \tag{15}$$

From the above equations:

- The term $Q(t', t'')$ is referred to the influence of the source/fiber/detector system.

- The term $Q_0(t', t'')$ depends on the fiber and the power coupling from to the source to the fiber and from the fiber to the detector.

- The spatial coherence of the source related to the fiber modes is provided by $C_{\mu\mu'}$.

- $\chi_{\nu\nu'}$ is defined as $\chi_{\nu\nu'} = \int_{A_r} e^*_\nu(\bar{r})e_{\nu'}(\bar{r})d\bar{r}$. In the special case where the detector collects all the incident light $\chi_{\nu\nu'}$ $\delta_{\nu\nu'}$.

- The term $\langle h_{\mu\nu}^{*}(t-t')h_{\mu'\nu'}(t-t'')\rangle$ is referred to the fiber dispersion and to the mode coupling.

This last term, relative to the propagation along the fiber, is composed of two parts, one describing the independent propagation of modes $h_{\mu\nu}^{*}(t-t')$ and a second one describing the power coupling between modes $h_{\mu'\nu'}(t-t'')$. For analysing this term, it is required to consider the N coupled mode propagation equations (field amplitudes) in the frequency domain which refer to an N-mode multimode fiber. A detailed study of this analysis can be found in [54, 82].

Although Eq. (14) reveals a nonlinear relationship between the output and the input electrical signals being not possible to define a transfer function, under several conditions linearization is possible yielding to a linear system with impulse response $Q(t)$. This linear response is given by [81]:

$$P(t) = \int_{-\infty}^{\infty} S(t')Q(t - t', t - t')dt' \tag{16}$$

The impulse response terms of the fiber can then be found by inverse Fourier transforming the above matrix elements. Upon substitution in Eq. (16) it is found that $P(t)$ is composed of two terms $P(t) = P^{U}(t) + P^{C}(t)$ being $P^{U}(t)$ the optical power in absence of mode coupling and $P^{C}(t)$ the contribution of modal coupling. Moreover, both the coupled and uncoupled parts can be divided into a linear and a non-linear term, respectively. These non-linear terms will contribute to the harmonic distortion and intermodulation effects. Grouping the linear contributions of the uncoupled and the coupled parts, and comparing the power of the lineal part of the total power received (sum of contributions from the coupled and uncoupled parts) with the power of one of the sidebands of the electric modulating signal, it is possible to obtain the final overall RF transfer function, yielding Eq. (17). For a detailed description of the evaluation of both terms, see the works reported in [54, 82].

$$H(\Omega) = \sqrt{1 + \alpha_C^2} \cdot e^{-\frac{1}{2}\left(\frac{\beta_o^2 \Omega z}{\sigma_r}\right)^2} \cdot \cos\left(\frac{\beta_o^2 \Omega^2 z}{2} + \arctan(\alpha_C)\right) \cdot \sum_{m=1}^{M} 2m(C_{mm}\chi_{mm} + G_{mm})e^{-2\alpha_m z}e^{-j\Omega\tau_m z} \tag{17}$$

The expression of Eq. (17) provides a description of the main factors affecting the RF frequency response of a multimode fiber link and can be divided as the product of three terms of factors. From the left to the right, the first term is a low-pass frequency response which depends on the first order chromatic dispersion parameter β_o^2 which is assumed to be equal for all the modes guided by the fiber, and the parameter σ_C which is the source coherence time directly related to the source linewidth. The second term is related to the Carrier Suppression Effect (CSE) due to the phase offset between the upper and lower modulation side-bands, as the optical signal travels along a dispersive waveguide, i.e. optical fiber. When the value of this relative phase offset is 180 degrees, a fading of the tone takes place. Finally, the third term represents a microwave photonic transversal filtering effect [83], in which each sample corresponds to a different mode group m carried by the fiber. Coefficients C_{mm}, X_{mm} and G_{mm} stand for the light injection efficiency, the mode spatial profile impinging the detector area and the mode coupling coefficient, respectively. This last term involves that the periodic frequency response of transversal filters could permit broadband RF, microwave and mm-wave transmissions far from baseband thus achieving a transmission capacity increase in such fiber links. Parameters α_{mm} and τ_{mm} represent the differential mode attenuation (DMA) effect and the delay time of the guided modes per unit length, respectively.

5. Analysis and results on silica-based multimode optical fibers

The MMF transfer function presented in Eq. (17) provides a description of the main factors affecting the RF frequency response of a multimode fiber link, including the temporal and spatial source coherence, the source chirp, chromatic and modal dispersion, mode coupling (MC), signal coupling to modes at the input of the fiber, coupling between the output signal from the fiber and the detector area, and the differential mode attenuation (DMA). Theoretical simulations and experimental results are studied with regards to several parameters in order to determine the optimal conditions for a higher transmission bandwidth in baseband and to investigate the potencials for broadband Radio-over-Fiber (RoF) systems in regions far from baseband using multimode fiber.

For the simulation results in this section it has been considered a 62.5/125µm core/cladding diameter graded-index multimode fiber (GI-MMF) with a typically SiO_2 core doped with 6.3 mol-% GeO_2 and a SiO_2 cladding, and intrinsic attenuation of 0.55dB/km. This typical doping value has been provided by the manufacturer. The refractive indices were approximated using a three-term Sellmeier function for 1300nm and 1550nm wavelengths. Sellmeier coefficients were provided by the manfacturer. Core and cladding refractive indices as a function of wavelength, from the Sellemier equation, Eq. (3), are illustrated in Fig. 9. A comparison of the core refractive index for a different core doped multimode fiber consisting of 7.5mol-% GeO_2 is given. The parameters relative to the differential mode attenuation were fitted to $\rho = 9$ and $\eta = 7.35$. Coefficient G_{mm} was obtained assuming a random coupling process defined by a Gaussian autocorrelation function [28] with a rms deviation of $\sigma = 0.0009$ and a correlation length of $\varsigma = 115 \cdot a$, being a the fiber core radius. The rms linewidth of the source was set to 10MHz and its chirp parameter to zero. A refractive index profile of $\alpha = 2$ was considered. Overfilled launching condition (OFL) was also assumed so that the light injection coefficient was set to $C_{mm} = 1 / M$, being M the total number of mode groups.

Fig. 10 illustrates the frequency response of a 3km-long GI-MMF link in absence and presence of DMA and mode coupling effects. An optical source operating at 1300nm and with

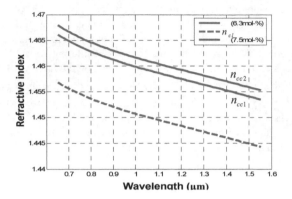

Figure 9. MMF core (n_{cci}) and cladding (n_{cl}) refractive index for different dopant concentration.

10MHz of linewidth has been considered. The filtering effect caused by the DMA is decreased when considering the presence of the mode coupling phenomenon. Moreover, the RF baseband bandwidth is increased by mode coupling while DMA has little effect on the bandwidth itself. Anyway, not considering mode coupling effects, Fig. 10 illustrates the classical conflict relationship between dispersion and loss in MMFs in general. As a matter of fact, the large DMA of high-order modes necessarily causes a large power penalty during light propagation, but at the same time it yields a bandwidth enhancement as a result of the mode stripping effect. Finally if mode coupling effects are considered, there is no deviation on the resonance central frequencies no matter the fiber DMA whilst DMA has a significant effect when mode coupling is considered to be negligible. $f_o \mid_n$ refers to the possible transmission channels far from baseband that could be employed.

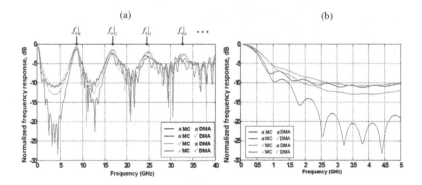

Figure 10. (a) Frequency responses up to 40GHz for a 62.5/125μm GIMMF showing the effect of mode coupling and DMA. L=3km. (b) Zoom up to 5GHz.

The influence of the optical fiber properties over its frequency response is of great importance. Parameters such as the core radius, the graded-index exponent, length and the core
refractive index count for this matter. Nevertheless, the most critical parameter that define
the behaviour and performance of a graded-index optical fiber type is its refractive index
profile α. It should be outlined that this index profile may slightly vary with wavelength,
always due to the eventually nonlinear Sellmeier coefficients. As a consequence of this, a
profile conceived to be optimal (in terms of bandwidth, for example) at a given wavelength
may will be far from optimal at another wavelength. This fact was also addressed in Section
2.2. The α-dispersion is imposed by the dopant and its concentration, so this impairment is
not easy to overcome. Furthermore, these latter parameters can also be affected by temperature impairments, as recently reported in [84]. Frequency responses are displayed in Fig.
11(a) for a 2km-long GI-MMF link showing the influence of 1% fiber refractive index profile
deviations on the RF transfer function. The rest of parameters for the simulations take the
same value as aforementioned. Significant displacements of the high-order resonances over
the frequency spectrum are noticed. From simulation conditions, attending to Fig. 11(a), an
increase of $\alpha' = \alpha + 0.04$ produces a change of the first-order resonance central frequency of
3.2GHz. It is also noticeable that the 3-dB passband bandwidth of the high-order resonances
is also highly influenced. Both facts could cause a serious MMF link fault if multiple-GHz
carriers are intended to be transmitted through this physical medium when performing a
RoMMF system.

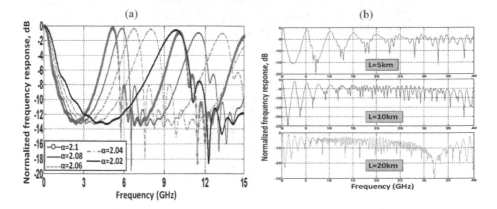

Figure 11. (a) Influence of the refractive index profile on the GI-MMF frequency response for a 2km-long link. (b) GI-
MMF frequency response for different link lengths, covering access reach.

The MMF frequency response dependence on the link length is shown in Fig. 11(b), covering
typical access network distances. High-order resonances far from baseband are slightly dis-
placed over the frequency spectrum with changes in attenuation depending on the case. In
addition, transmission regions can be easily identified as well as the effect of the carrier sup-

pression (CSE) due to the presence of intermediate notches, as seen in the case of L=20km. This effect can not be overlooked but could be avoided using single sideband modulation

Finally, the following figures illustrate both the influence of the optical source linewidth characteristic as well as the launching condition with regards to the GI-MMF frequency response. The influence of other optical source characterisitics such as the source chirp and the operating wavelength can be seen in [54]. It should be noted that wavelength emission provided by the optical source links with different optical fiber properties to be considered. Parameters such as the core and cladding refractive indices, the material dispersion, the propagation constant, the intrinsic attenuation and the number of propagated modes strongly depend on the optical wavelength launched into the fiber, being not an easy task to determine a real comparison about the influence of this parameter on the frequency response.

Fig. 12(a) illustrates the GI-MMF frequency response of a 3km-long link for three different optical sources operating at 1300nm. The rest of parameters take the same value as those previously indicated. The response for the DFB laser (with a Full Width Half Maximum - FWHM- of 10MHz) behaves relatively flat at high frequencies. The frequency response employing a FP laser with 5.5nm linewidth however suffers from a low-pass effect, determined by a 40dB fall at 40GHz. In the case of using a broadband light source, such a Light Emitting Diode (LED) with 30nm of source linewidth, the response falls dramatically after a few GHz and no high-order resonances are observed. On the other hand, the influence of the launching condition on the frequency response can be seen in Fig. 12(b). A GI-MMF link of 1km and an input power spectral density conforming a Gaussian lineshape from a DFB optical source with 10MHz FWHM have been considered. From the frequency response it is noticeable the dramatic enhancement of the baseband bandwidth as well as the achievement of a flat response in all the 20GHz-spectrum considered.

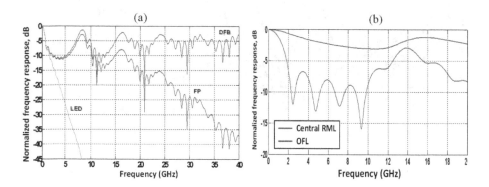

Figure 12. (a) Influence of the optical source temporal coherence on the frequency response of a 3km-long GI-MMF link. (b) Influence of the light injection on the frequency response of a 1km-long GI-MMF link.

By evaluating the latter results, it is observed that exploiting the possibility of transmitting broadband signals at high frequencies is contingent on the use of both narrow-linewidth optical sources and selective mode-launching schemes. These requirements were also confirmed in the work reported in [82].

Some measurement examples of the silica-based MMF transfer function are presented highlighting the conditions upon broadband MMF transmission in regions far from baseband can be featured thus validating the theoretical model described and proposed in [81]. The setup schematic for the experimental measurements is shown in Fig. 13.

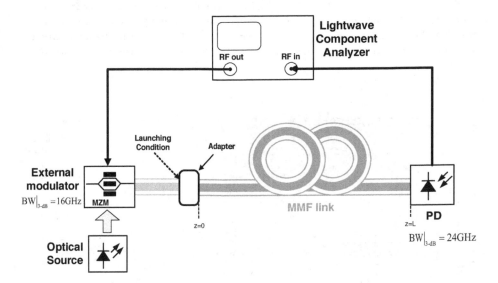

Figure 13. Block diagram of the experimental setup for the silica-based GI-MMF frequency response measurement up to 20GHz.

A Lightwave Component Analyzer (LCA, Agilent 8703B, 50MHz–20GHz) has been used to measure the frequency response, using a 100Hz internal filter. In all cases the laser was externally intensity modulated with an RF sinusoidal signal up to 20GHz of modulation bandwidth, by means of an electro-optic (E/O) Mach-Zehnder modulator (model JDSU AM-130@1300nm and JDSU AM-155@1550nm). At the receiver, the frequency response is detected by using a high-speed PIN photodiode, model DSC30S, from Discovery Semiconductors. It should be mentioned that the experimental results of the silica-based GI-MMF link shown in this section have been calibrated with regards to both the E/O intensity modulator and the photodetector electrical responses, being therefore solely attributed to the MMF fiber. It should be also noted that the ripples observed are caused by reflections in the optical system and are not features of the fiber response, although FC/APC connectors are used to minimize this effect. To perform different launching conditions, the optical output of the E/O modulator was passed through a 62.5/125µm silica-based MMF fiber patch cord plus a mode scrambler before being launched to the MMF link. This optical launching scheme provides an OFL condition for light injection. On the other hand, selective central mode launching was achieved by injecting light to the system via a SMF patchcord.

Fig. 14(a) shows the measured frequency response for a 3km silica-based GI-MMF link. As it was expected from the theory, while the response for the DFB laser (@1550nm) behaves relatively flat at high frequencies, with maximum variations of approximately ± 0.8 dB with regards to a mean level of approximately 2.5dB below the low frequency regime, the response relative to the FP laser (@1310nm) suffers from a low pass effect characterized by a 15dB fall at 20GHz. In the case of the Broadband Light Source (BLS), the response falls dramatically after a few GHz. Therefore, as previously stated, exploiting the possibility of transmitting broadband RF signals at high frequencies is contingent on the use of narrow-linewidth sources. This latter performance stands regardless the operating wavelength from the optical source.

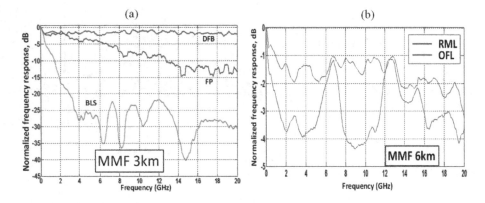

Figure 14. (a) Measured influence of the optical source linewidth on the silica-based GI-MMF frequency response. (b) Measured influence of the launching condition on the silica-based GI-MMF frequency response.

Additionally, two launching conditions, RML and OFL, were also applied to the fiber link. Results are shown in Fig. 14(b), and were performed by using a DFB laser operating at with FWHM of 100kHz. As expected, for the RML condition, in which a limited number of modes is excited, the typical transversal filtering effect of the MMF is significantly reduced, thus achieving an increased flat response over a broader frequency range spectrum. It should be noted that the distance values comprising Fig. 14 are representative of currently deployed moderate-length fiber links.

Finally, the above figures show a comparison between the curves predicted by the theoretical model and the experimental results showing good agreement between them. A FP source operating at 1300nm with $\Delta\lambda=1.8$nm of source linewidth has been employed in measurements reported in Fig. 15(a). An OFL excitation at the fiber input end has been applied. Theoretical curves have been obtained considering a silica-based MMF with a SiO_2 core doped with 6.3mol-% GeO_2 and a SiO_2 cladding, with a refractive index profile of

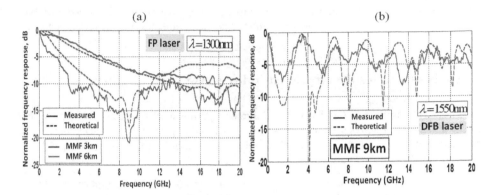

Figure 15. (a) Theoretical and measured frequency response of a 3km- and 6km-long silica based GI-MMF link with a FP laser source operating at 1300nm. (b) Theoretical and measured frequency response of a 9km-long silica-based GI-MMF link with a DFB laser source operating at 1550nm [84].

α =1.921 and an intrinsic attenuation coefficient of α_0=0.59dB/kmat 1300nm. The latter was measured employing Optical Time-Domain Reflectometer (OTDR) techniques. Core and cladding refractive indices have been calculated using a three-term Sellmeier function. It has also been assumed a free chirp source. Differential Mode Attenuation (DMA) effects have been considered by setting ρ =8.7; η =7.35. Additionally, a random coupling process defined by a Gaussian autocorrelation function has beeen defined for the mode coupling with a correlation length of ς =0.0036m and rms deviations of σ =0.0012@3km and σ =0.0017@6km. Fig. 15(a) also addresses the high-order resonances (passband) suppression effect as the source linewidth increases.. This is due to the fact that in this latter case the low pass term in Eq. (17) begins to dominate over the other two. In contrast, in Fig. 15(b), a DFB laser source with 100kHz of linewidth and operating at 1550nm has been employed. An intrinsic fiber attenuation coefficient of α_0=0.31dB/km at 1550nm was measured and a rms deviation of σ =0.0022@9km was considered for a link length of 9km. The rest of parameters take the same values as aforementioned. Several passband channels suitable for multiple-GHz carrier transmission over the frequency spectrum are observed as well as a relatively flat region over 17GHz. However, a significant discrepancy can be observed in the resonances excursions, being the measured ones not so pronounced compare to what the model predicts, i.e. the measured filtering effect is decreased compare to what it is expected. Many reasons can be attributed for this behaviour but mainly due to both the DMA and mode coupling modelling approaches considered.

Finally, although the 3-dB bandwidth of the baseband frequency response has not been paid much attention in this analysis, it is commonly agreed that the measurement uncertainty in characterizing this parameter is quite large and a standard deviation on the order of 10%-20% or more is not uncommon. This performance depends on the care of a particular lab in setting up the launch conditions and acquiring the data. This was verified in 1997 with an informal industry wide round robin [85]. Furthermore, it was, in fact, because of this that the industry standardized the overfilled launch (OFL) condition during the late 1990s [86].

6. Analysis and results on graded-index polymer optical fibers

This section, comprising Graded-Index Polymer Optical Fibers (GIPOFs) will follow the same scheme as previous section. Furthermore, this section proves that the same principles are essentially valid for silica-based MMFs and GIPOFs in order to extend their capabilities beyond the RF baseband bandwidth.

For the simulation results in this section it has been considered a 120/490μm core/cladding diameter graded-index polymer optical fiber (PF GIPOF) with intrinsic attenuation of 60dB/km at 1300nm and 150dB/km at 1550nm. The refractive indices for the fiber core and fiber cladding were calculated using a three-term Sellmeier. These coefficients were provided by the manfacturer. Core and cladding refractive indices as a function of wavelength, from the Sellemier equation, Eq. (3), are illustrated in Fig. 16. The parameters relative to the differential mode attenuation were fitted to $\rho=11$ and $\eta=12.2$. Coefficient G_{mm} was obtained assuming a random coupling process defined by a Gaussian autocorrelation function with a rms deviation of $\sigma=0.0005$ and a correlation length of $\varsigma=1.6\times10^4\cdot a$, being a the fiber core radius. This latter value of the correlation length provides similar mode coupling strengths than that of reported in other works for PF GIPOF fibers such as in [35, 44]. The rms linewidth of the optical source was set to 5nm and its chirp parameter to zero. A refractive index profile of $\alpha=2$ was considered, unless specified. Overfilled launching condition (OFL) was also assumed so that the light injection coefficient was set to $C_{mm}=1/M$, being M the total number of mode groups.

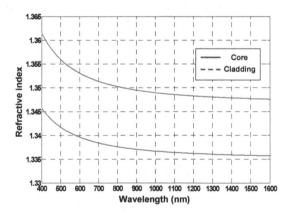

Figure 16. PF GIPOF core and cladding refractive index as a function of wavelength.

It is worth mentioning that with PF GIPOFs, a thermally determined alteration in the dopant material can come about, leading to changes in the refractive index, although new materials have just recently become available and behave with admirable stability in this issue [87]. This dopant concentration during the manufacturing process is also directly related to the fiber refractive index profile.

Fig. 17(a) depicts the frequency response of a 200m-long PF GIPOF link operating at 1300nm in absence and presence of DMA and mode coupling effects. The theoretical curve for a 200m-long PMMA GIPOF in presence of both DMA and mode coupling is also given for comparison. As expected, much greater baseband bandwidths are obtained by using fluorine dopants in the core instead of classical PMMA-based composites. The results indicate that the presence of both effects is favorable for improving the frequency response of the GIPOF. It can be observed a more than a three-fold RF baseband bandwidth enhancement caused if only DMA effect is considered. This result shows that the DMA is a determining factor for accurate assessment of the baseband in GIPOFs. No high-order resonances are shown due to both the high fiber attenuation and the OFL launching condition considered. As in the case of silica-based MMFs, the influence of the optical fiber properties over its frequency response is of great importance. Parameters such as the core radius, the graded-index exponent, length and the core refractive index count for this matter. Similar mechanisms as those stated for silica-based GI-MMFs rule for PF GIPOFs concerning these parameters. This fact is illustrated in Fig. 17(b), in which PF GIPOF frequency responses are displayed for a 200m-long link showing the influence of 5% fiber refractive index profile deviations on the RF transfer function. The rest of parameters for the simulations take the same value as aforementioned. Similarly to silica-based counterparts, significant displacements of the high-order resonances over the frequency spectrum are noticed. However, it is worth pointing out that PF GIPOFs are less sensitive to α-tolarences compared to that of silica counterparts.

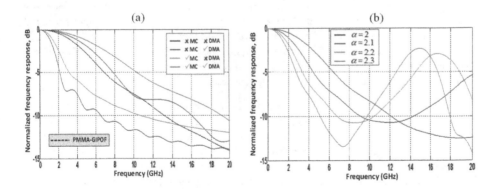

Figure 17. (a) Frequency responses up to 20GHz for a 200m-long PF GIPOF link showing the effect of mode coupling and DMA. Similar PMMA-based link is also illustrated for comparison. (b) Influence of the refractive index profile on the PF GIPOF frequency response for a 200m-long link.

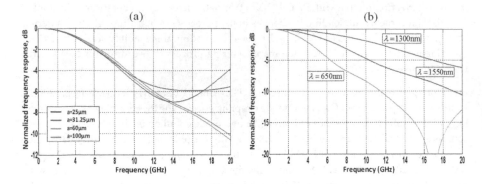

Figure 18. (a) Influence of the core radius on the PF GIPOF frequency response at OFL condition. (b) Influence of the operating wavelength on the PF GIPOF frequency response.

On the other hand, Fig. 18(a) illustrates the frequency response of present commercially available PF GIPOFs with different core radius. Identical simulation parameters have been considered. From the theoretical curves, similar RF baseband bandwidths at OFL condition are obtained, independent from the core radius considered, although high-order resonances start to notice as core radius decreases. However, this fact turns to be different if RML launching is applied. Simulations under this light injection condition predict that lower fiber core radius results in a RF baseband bandwidth enhancement. This result is quite in agreement with the fact that the bandwidth reduction is to be connected with the larger number of excited modes, directly related to the fiber core radius. Nevertheless, this dependence is strongly reduced as nearer OFL is reached. Moreover, the frequency response dependence on the operating wavelength is shown in Fig. 18(b). As expected, due to the high chromatic dispersion of PF GIPOFs at 650nm, see Fig. 3(a), RF baseband bandwidth at this wavelength falls dramatically after few GHz. On the other hand, baseband bandwidths achievable at 1300nm are greater than those obtained at 1550nm despite the similar PF GIPOF material dispersion (even slight smaller at 1550nm) and despite the use of a relatively narrow-linewidth optical source. Thus, bandwidth must mostly be limited by modal dispersion. The reason for this bandwidth difference is supported by the fact that DMA effects are supposed to be stronger at 1300nm than that of 1550nm, leading to a RF baseband bandwidth enhancement.

Finally, the following figures illustrate both the influence of the optical source linewidth characteristic as well as the launching condition with regards to the PF GIPOF frequency response. The influence of other optical source characterisitics such as the source chirp and the operating wavelength can be seen in [54]. Fig. 19(a) illustrates the PF GIPOF frequency response at 1300nm of a 200m-long link for: a DFB optical source with 10MHz of FWHM; a FP laser of 2nm of linewidth; and a LED with 20nm of source linewidth. The rest of parameters take the same value as those previously indicated. As expected, the frequency response is progressively penalised as source linewidth increases, hampering the possible observance of high-order resonances. On the other hand, the influence of the launching condition on the frequency response can be seen in Fig. 19(b). A PF GIPOF link of 200m and an input power spectral density conforming a Gaussian lineshape from a DFB optical source with 0.2nm have been considered. From the frequency response, a dramatic enhancement of the RF baseband bandwidth is observed when applying a RML condition as well as a reduction of the filtering effect, similarly of what it was expected from the silica-based MMF analysis.

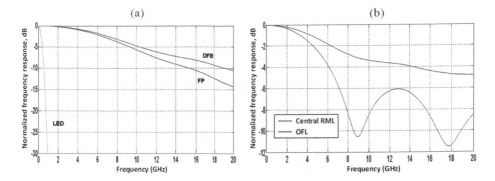

Figure 19. (a) Influence of the optical source temporal coherence on the frequency response of a 200m-long PF GIPOF link. (b) Influence of the light injection on the frequency response of a 200m-long PF GIPOF link.

Some measurement examples of the PF GIPOF transfer function are presented highlighting the conditions upon broadband transmission in regions far from baseband can be featured thus validating the theoretical model proposed [88]. A comparison between the curves predicted by the theoretical model and the experimental results is also provided. Good agreement between theory and experimental results is observed. The results have been tested over an amorphous perfluorinated (PF) graded-index polymer optical fiber. In all cases, such fiber type fulfils the requirements of the IEC[6] 60793-2-40 standard for the PF polymer-based POFs (types A4f, A4g and A4h) which fits a minimum bandwidth of 1500MHz@100m for A4f type and 1880Mhz@100m for A4g and A4h types, respectively. The setup schematic for the experimental measurements follows the same concept as that reported in Fig. 13. The experimental results have been calibrated with regards to both the E/O intensity modulator and the photodetector electrical responses. Similar optical sources as those used in silica-based MMFs experiments were employed. An OFL excitation at the fiber input end has been applied. Theoretical curves have been obtained considering a PF GIPOF with a refractive index profile of $\alpha = 2.18$ and an intrinsic attenuation coefficient of $\alpha_o = 42$dB/km at 1300nm. The latter was measured employing Optical Time-Domain Reflectometer (OTDR) techniques. Core and cladding refractive indices have been calculated using a three-term Sellmeier function. It has also been assumed a free chirp source. Differential Mode Attenuation (DMA) effects have been considered by setting $\rho = 11$; $\eta = 12.2$. Additionally, a random coupling process defined by a Gaussian autocorrelation function has beeen defined for the mode cou-

---

6 International Electrotechnical Commission

pling with a correlation length of $\varsigma = 0.005$m and rms deviation of $\varsigma = 1.6 \cdot 10^4 \times$a, being 'a' the core radius of the fiber considered.

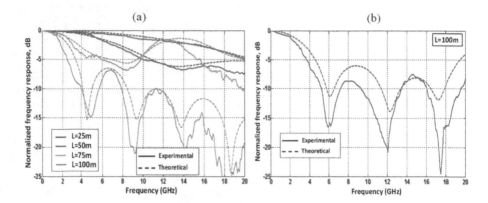

Figure 20. (a) Theoretical and measured frequency response of a 50μm core diameter PF GIPOF link for different lengths with a FP laser source operating at 1300nm. (b) Theoretical and measured frequency response of a 100m-long 62.5μm core diameter PF GIPOF link under identical operating conditions.

Fig. 20(a) depicts the theoretical (dashed line) and measured (solid line) frequency responses of a 50μm core diameter PF GIPOF link for different lengths. On the other hand, the theoretical and measured frequency response of a 100m-long 62.5μm core diameter PF GIPOF link is shown in Fig. 20(b). In both cases, a FP optical source operating at 1300nm and 1.8nm of linewidth was employed. Results reveal the presence of some latent high-order resonances in the PF GIPOF frequency response. Although these passbands suffer from a power penalty in the range of 5dB per passband order, attending to Fig. 20(a), this high attenuation could significantly be improved with lower fiber attenuation values. Nevertheless, the presence of these periodicities in the PF GIPOF frequency response opens up the extension of the transmission capabilities beyond baseband thus increasing the aggregated capacity over this optical fiber type.

Another example can be seen in Fig. 21(a), where the theoretical and measured frequency response of a 150m-long 120μm core diameter PF GIPOF link is depicted. Similar operating conditions as above were applied. In constrast, Fig. 21(b) reports the RF bandwidth enhancement when employing a narrow-linewidth DFB optical source. A 62.5μm core diameter PF GIPOF was used. As expected, the available bandwidth is increased if we compare the curves within this figure with those obtained in Fig. 20(a). However, it is important to observe that the frequency response at 1550nm falls at 17dB at 20GHz. This is due to the fact that the PF GIPOF intrinsic attenuation at this wavelength was measured to be 140dB/km. In both figures an OFL condition was also applied.

Finally, the following figure evaluates the conditions upon which broadband transmission over PF GIPOF beyond the RF baseband bandwidth is possible. Fig. 22(a) shows the meas-

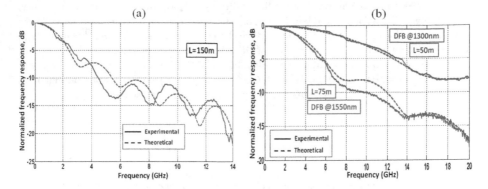

Figure 21. (a) Theoretical and measured frequency response of a 150m-long 120μm core diameter PF GIPOF link, under similar operation conditions as in Fig. 20. (b) Theoretical and measured frequency response of a 62.5μm core diameter PF GIPOF link employing DFB optical sources.

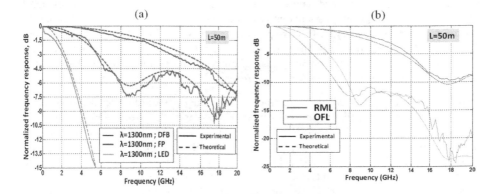

Figure 22. (a) Influence of the optical source temporal coherence on the frequency response of a 50m-long 62.5μm core diameter PF GIPOF link. (b) Measured influence of the launching condition on the 50m-long 120μm core diameter PF GIPOF frequency response.

ured frequency response for a 50m-long 62.5μm core diameter PF GIPOF link at an operating wavelength of 1300nm. As it was expected from the theory, the frequency response dramatically decreases when increasing the rms source linewidth. When a LED with $W = 98$nm of spectral width is employed as the optical source, the frequency response falls after a few GHz. In contrast greater baseband bandwidths when employing a FP laser with $W = 1.8$nm and a DFB source with 100kHz of FWHM are achievable. In addition, the presence of high-order resonances in the frequency response is also identified. On the other hand, Fig. 22(b) illustrates the frequency response of a 50m-long 120μm core diameter PF GIPOF link at launching conditions OFL and RML, respectively. In both cases a FP laser operating at 1300nm and 1.8nm of source linewidth was employed. As expected from theory, RML increases the RF baseband bandwidth as well as flattens the frequency response. However, possible transmission regions beyond baseband are penalised in power due to the high PF GIPOF attenuation compared to that of silica-based counterparts. From both figures, we can therefore conclude, and similarly to silica-based MMFs, that exploiting the possibility of transmitting broadband RF signals in PF GIPOFs at high frequencies is also contingent on the use of narrow-linewidth sources and selective mode-launching schemes.

7. Conclusions

Future Internet Access technologies are supposed to bring us a very performing connection to the main door of our homes. At the same tine, new services and devices and their increase use, commonly grouped as next-generation access (NGA) services, will require data transfers at speeds exceeding 1Gbps inside the building or home at the horizon 2015. Both drivers lead to the deployment of a high-quality, future-proof network from the access domain even to inside buildings and homes. There is a widely-spread consensus that FTTx is the most powerful and future-proof access network architecture for providing broadband services to residential users. Furthermore, FTTx deployments with WDM-PON topologies are considered in the long-term the target architecture for the next-generation access networks. This environment may end up taking advantage of optical cabling solutions as an alternative to more traditional copper or pure wireless approaches.

Multimode optical fibers (MMF), both silica- and polymer-based, can offer the physical infrastructure to create a fusion and convergence of the access network via FFTx for these next-generation access (NGA) services. Both fiber types may be used not only to transport fixed data services but also to transparently distribute in-building (and also for short- and medium-reach links) widely ranging signal characteristics of present and future broadband services leading to a significant system-wide cost reduction. The underlying reason is that multimode fibers have a much larger core diameter and thus alignment in fiber splicing and connectorising is more relaxed. Also the injection of light from optical sources is easier, without requiring sophisticated lens coupling systems. And these facts seem to be critical as all optical networks are being deployed even closer to the end users, where most of interconnections are needed. Moreover, polymer optical fiber (POF) may be even easier to install than silica-based multimode fiber, as it is more ductile, easier to splice and to connect even maintaining high bandwidth performances as in the case of PF GIPOFs. However, their main drawback is related to the fact that their bandwidth per unit length is considerably less with respect to singlemode fiber counterparts. However, this may not be decisive as link lengths are relatively short in the user environment.

On the other hand, it is obvious that the deployment of such emerging NGA network technology and its convergence would be not possible without the research and evaluation of predictive and accurate models to describe the signal propagation through both MMF fiber types to overcome the inherent limitations of such a transport information media. However, the potentials of MMF to support broadband RF, microwave and millimetre-wave transmission beyond baseband over short and intermediate distances are yet to be fully known, as its frequency response seems to be unpredictable under arbitrary operating conditions as well as fiber characteristics. The different experimental characterizations and the theoretical model presented in this chapter allow understanding and an estimation of the frequency response and the total baseband bandwidth. In addition, can give an estimation of the aggregated transmission capacity over MMFs through analyzing the high-order resonances as well as the presence of relatively flat regions that are present beyond baseband, under certain conditions, in the MMF frequency response.

From the theoretical and experimental results, it is demonstrated that, next to its baseband transmission characteristics, an intrinsically multimode fiber link will show passband characteristics in higher frequency bands. However, the location and the shape of these passbands depend on the actual fiber characteristics, and may change due to environmental conditions and/or the light launching conditions as well as the number of guided modes excited and the power distribution among them. Also the length of the fiber, the mode coupling processes, the source wavelength, the launching scheme, and the fiber core diameter influence the fibre frequency response. This fact imposes a great challenge for the extension of the bandwidth-dependent multimode fiber performance. And, the influence of most of all these parameters that can have a large impact on the date rate transmission performance in MMF links has been addressed. Although no accurate agreement can be expected due to the many approximations made in the theoretical analysis as well as the amount of parameters involved in the frequency response, the results reveal a quite good assessment in the behavior of the multimode optical fiber frequency response compared to the curves predicted by the model

The use of selective mode-launching schemes combined with the use of narrow-linewidth optical sources is demonstrated to enable broadband RF, microwave and millimetre-wave transmission overcoming the typical MMF bandwidth per length product. Under these conditions it is possible to achieve flat regions in the frequency response as well as passband characteristics far from baseband. Transmission of multiple-GHz carrier in these MMF links can be featured at certain frequencies albeit a small power penalty, enabling the extension of broadband transmission, with direct application in Radio-over-Fiber (RoF) systems, in which broadband wireless services could be integrated on the same fiber infrastructure, thereby reducing system costs. The results also reveal that PF GIPOF has some latent high-order passbands and flat regions in its frequency response, which however suffer a high attenuation due to the higher intrinsic attenuation of polymer optical fibers compared to that of silica-based counterparts. Anyway, this power penalty could significantly be improved with lower fiber attenuation values.

To resume, MMFs (both silica- and polymer-based) are still far from SMF bandwidth and attenuation, but they are called to the next step on access network links due to its low cost systems requirements (light sources, optical detectors, larger fiber core,...) against the high cost of the singlemode components. It is worth mentioning that in-building networks may comprise quite a diversity of networks: not only networks within residential homes, but also networks inside office buildings, hospitals, and even more extensive ones such as networks in airport departure buildings and shopping malls. Thus the reach of in-building networks may range from less than 100 metres up to a few kilometers. A better understanding of the possibilities of signal transmission outside the baseband of such fibers are investigated, in order to extend their capabilities, together with the evaluation of current fiber frequency response theoretical models becomes of great importance.

Acknowledgements

The work comprised in this document has been developed in the framework of the activities carried out in the Displays and Photonics Applications group (GDAF) at Carlos III University of Madrid.

This research work has been supported by the following Spanish projects: TEC2006-13273-C03-03-MIC, TEC2009-14718-C03-03 and TEC2012-37983-C03-02 of the Spanish Interministerial Comission of Science and Technology (CICyT) and FACTOTEM-CM:S-0505/ESP/000417 and FACTOTEM-2/2010/00068/001 of Comunidad Autónoma de Madrid.

Additional financial support was obtained from the European Networks of Excellence: ePhoton/One+ (FP6-IST-027497)[7] and and BONE: Building the Future Optical Network in Europe (FP7-ICT-216863)[8]

Author details

David R. Sánchez Montero and Carmen Vázquez García

Displays and Photonics Applications Group (GDAF), Electronics Technology Dpt., Carlos III University of Madrid, Leganés (Madrid), Spain

References

[1] Broadband Network Strategies; Strategy Analytics. Global Broadband Forecast: 2H2011 18 Nov 2011.

[2] Charbonnier, B. End-user Future Services in Access, Mobile and In-Building Networks. FP7 ICT-ALPHA project, public deliv. 1.1p, http://www.ict-alpha.eu, July 2008.

---

7 ePhoton/One+ is supported by the Sixth Framework Programme (FP6) of the European Union.
8 BONE is supported by the Seventh Framework Programme (FP7) of the European Union.

[3]   Meyer, S. Final usage Scenarios Report. FP7 ICT-OMEGA Project, public deliv. 1.1 Aug. 2008.

[4]   Ishigure, T., Aruga, Y. and Koike Y. High-bandwidth PVDF-clad GI POF with ultra-low bending loss. Journal of Lightwave Technology 2007; 25(1) 335-345.

[5]   Ishigure, T., Makino, K., Tanaka, S. and Koike, Y. High-Bandwidth Graded-Index Polymer Optical Fiber Enabling Power Penalty-Free Gigabit Data Transmission. Journal of Lightwave Technology 2003; 21(11) 2923-2930.

[6]   Ziemann, O., Krauser J., Zamzow, P.E. and Daum, W. POF Handbook: Optical Short Range Transmission Systems. Berlin: Springer 2nd Edition; 2008.

[7]   Epworth, R.E. The phenomenon of modal noise in analog and digital optical fiber systems. Proceedings of the 4th European Conference on Optical Communications (ECOC), Genova, Italy; 1978.

[8]   Bates, R.J.S., Kutcha, D.M. Improved multimode fibre link BER calculations due to modal noise and non-self-pulsating laser diodes. Optical and Quantum Electronics 1995; 27 203-224.

[9]   Pepeljugoski, P., Kuchta, D.M. and Risteski, A. Modal noise BER calculations in 10-Gb/s multimode fiber LAN links. IEEE Photonics Technology Letters 2005; 17(12) 2586-2588.

[10] Gasulla I. and Capmany, J. Modal noise impact in Radio over Fiber multimode fiber links. Optics Express 2008; 16(1) 121-126.

[11] Daum, W. Krauser, J., Zamzow P. and Ziemann, O. POF - Polymer Optical Fibers for Data Communication. Germany: Springer; 2002.

[12] Li, W., Khoe, G.D., van den Boom, H.P.A., Yabre, G., de Waardt, H., Koike, Y., Yamazaki, S., Nakamura, K. and Kawaharada, Y. 2.5 Gbit/s transmission over 200 m PPMA graded index polymer optical fiber using a 645 nm narrow spectrum laser and a silicon APD. Microwave and Optical Technology Letters 1999; 20(3) 163-166.

[13] Groh, W. Overtone absorption in macromolecules for polymer optical fibers. Makromoleculare Chemie 1988; 189 2861-2874.

[14] Ishigure, T., Nihei, E. and Koike, Y. Graded-index polymer optical fiber for high-speed data communication. Applied Optics 1994; 33(19) 4261-4266.

[15] Van den Boom, H.P.A., Li, W., van Bennekom, P.K., Tafur Monroy,I. and Khoe, G.D. High-capacity transmission over polymer optical fiber. IEEE Journal on Selected Topics in Quantum Electronics 2001; 7(3) 461-470.

[16] Nihei, E., Ishigure, T., Taniott, N. and Koike, Y. Present prospect of graded-Index plastic optical fiber in telecommunication. IEICE Transactions on Electronics 1997; E80-C 117-121.

[17] Tanio, N. and Koike, Y. What is the most transparent polymer?. Polymer Journal 2000; 32(1) 119-125.

[18] Ishigure, T., Tanaka, S., Kobayashi, E. and Koike, Y. Accurate refractive index profiling in a graded-index plastic optical fiber exceeding gigabit transmission rates. Journal of Lightwave Technology 2002; 20(8) 1449-1456.

[19] Gloge, D. Dispersion in Weakly Guiding Fibers. Applied Optics 1971; 10(11) 2442-2445.

[20] Liu, Y.E., Rahmna, B.M.A., Ning, Y.N. and Grattan, K.T.V. Accurate Mode Characterization of Graded-Index Multimode Fibers for the Application of Mode-Noise Analysis. Applied Optics 1995; 34 1540-1543.

[21] Xiao, J.B. and Sun, X.H. A Modified full-vectorial finite-difference beam propagation method based on H-fields for optical waveguides with step-index profiles. Optics Communications 2006; 266(2) 505-511.

[22] Huang, W.P. and Xu, C.L. Simulation of 3-Dimensional Optical Wave-Guides by a Full-Vector Beam-Propagation Method. IEEE Journal of Quantum Electronics 1993; 29(10) 2639-2649.

[23] Zubia, J., Poisel, H., Bunge, C.A., Aldabaldetreku, G. and Arrue, J. POF Modelling. Proceedings of the 11th International Conference on Plastic Optical Fiber (ICPOF) Tokyo, Japan, 221-224; 2002.

[24] Marcatili, E.A.J. Modal Dispersion in Optical Fibers with Arbitrary Numerical Aperture and Profile Dispersion. Bell Labs Technical Journal 1977; 6(1) 49-63.

[25] Ishigure, T., Nihei, E. and Koike, Y. Optimum refractive-index profile of the graded-index polymer optical fiber, toward gigabit data links. Applied Optics 1996; 35(12) 2048-2053.

[26] Olshansky, R. and Keck, D.B. Pulse broadening in graded-index optical fibers. Applied Optics 1976; 15 483-491, and errata; Soudagar M.K. and Wali, A.A. Applied Optics 1993; 32 6678.

[27] Olshansky, R. Mode-Coupling Effects in Graded-Index Optical Fibers. Applied Optics 1975; 14 935-945.

[28] Marcuse, D. Theory of Dielectric Optical Waveguides. New York: Academic Press 2nd Edition; 1991.

[29] Yabre, G. Comprehensive Theory of Dispersion in Graded-Index Optical Fibers. Journal of Lightwave Technology 2000; 18(2) 166-177.

[30] Presby, H.M. and Kaminow, I.P. Binary silica optical fibers: refractive index and profile dispersion measurements. Applied Optics 1976; 15(12) 469-470.

[31] Olshansky, R. and Nolan, D.A. Mode-dependent attenuation of optical fibers: Excess Loss. Applied Optics 1976; 15 1045-1047.

[32] Gloge, D. Weakly Guiding Fibers. Applied Optics 1971; 10(10) 2252-2258.

[33] Yabre, G. Theoretical investigation on the dispersion of graded-index polymer optical fibers. Journal of Lightwave Technology 2000; 18(6) 869-877.

[34] Mickelson, A.R. and Eriksrud, M. Mode-dependent attenuation in optical fibers. Journal of the Optical Society of America 1983; 73(10) 1282-1290.

[35] Golowich, S.E., White, W., Reed, W.A. and Knudsen, E. Quantitative estimates of mode coupling and differential modal attenuation in perfluorinated graded-index plastic optical fiber. Journal of Lightwave Technology 2003; 21(1) 111-121.

[36] White, W.R., Dueser, M., Reed, W.A. and Onishi, T. Intermodal dispersion and mode coupling in perfluorinated graded-index plastic optical fiber. IEEE Photonics Technology Letters 1999; 11(8) 997-999.

[37] Cancellieri, G. Mode-Coupling in Graded-Index Optical Fibers Due to Perturbation of the Index Profile. Applied Physics 1980; 23 99-105.

[38] Cancellieri, G. Mode-Coupling in Graded-Index Optical Fibers Due to Micro-Bending. Applied Physics and Materials Science & Processing 1981; 26 51-57.

[39] Su, L., Chiang, K.S. and Lu, C. Microbend-induced mode coupling in a graded-index multimode fiber. Applied Optics 2005; 44(34) 7394-7402.

[40] Ohashi, M., Kitayama, K. and Seikai, S. Mode-Coupling Effects in a Graded-Index Fiber Cable. Applied Optics 1981; 20 2433-2438.

[41] Djordjevich A. and Savovic, S. Investigation of mode coupling in step index plastic optical fibers using the power flow equation. IEEE Photonics Technology Letters 2000; 12(11) 1489-1491.

[42] Donlagic, D. Opportunities to enhance multimode fiber links by application of overfilled launch. Journal of Lightwave Technology 2005; 23(11) 3526-3540.

[43] Ishigure, T., Ohdoko, K., Ishiyama, Y. and Koike, Y. Mode-coupling control and new index profile of GI POF for restricted-launch condition in very-short-reach networks. Journal of Lightwave Technology 2005; 23(12) 4155-4168.

[44] Polley, A. High Performance Multimode Fiber Systems: A Comprehensive Approach. PhD Thesis. Georgia Institute of Technology; December 2008.

[45] Cunningham, D.G., Lane, W.G. and LAne, B. Gigabit Ethernet Networking. Indianapolis: Macmillan; 1999.

[46] Gasulla I. and Capmany, J. 1 Tb/s km Multimode fiber link combining WDM transmission and low-linewidth lasers. Optics Express 2008; 16(11) 8033-8038.

[47] Sim, D.H., Takushima, Y. and Chung Y.C. High-Speed Multimode Fiber Transmission by Using Mode-Field Matched Center-Launching Technique. Journal of Lightwave Technology 2009; 27(8) 1018-1026.

[48] Zeng, J., Lee, S.C.J., Breyer, F., Randel, S., Yang, Y., van den Boom, H.P.A. and Koonen, A.M.J. Transmission of 1.25 Gb/s per Channel over 4.4. km Silica Multimode Fibre Using QAM Subcarrier Multiplexing. Proceedings of the 33th European Conference on Optical Communications (ECOC), paper 7.4.3, Berlin, Germany; 2007.

[49] WiMedia Alliance. Worlwide regulatory status. http://www.wimedia.org.

[50] Koonen, A.M.J. and Garcia Larrode, M. Radio-Over-MMF Techniques, Part II: Microwave to Millimeter-Wave Systems. Journal of Lightwave Technology 2008; 26(15) 2396-2408.

[51] Lee, S.C.J., Breyer, F., Randel, S., Spinnler, B., Polo, I.L.L., van Den Borne, D., Zeng, J., de Man, E., van Den Boom, H.P.A. and Koonen, A.M.J. 10.7 Gbit/s transmission over 220 m polymer optical fiber using maximum likelihood sequence estimation. Proceedings of Optical Fiber Communication and the National Fiber Optic Engineers Conference (OFC/NFOEC), Anaheim, USA; 2007.

[52] Polley, A. and Ralph, S.E. 100 m, 40 Gb/s Plastic Optical Fiber Link. Proceedings of Optical Fiber Communication and the National Fiber Optic Engineers Conference (OFC/NFOEC), San Diego, USA; 2008.

[53] Yang, H., Lee, S.C.J., Tangdiongga, E., Okonkwo, C., van den Boom, H.P.A., Breyer, F., Randel, S. and Koonen, A.M.J. 47.4 Gb/s Transmission Over 100 m Graded-Index Plastic Optical Fiber Based on Rate-Adaptive Discrete Multitone Modulation. Journal of Lightwave Technology 2010; vol. 28(4) 352-359.

[54] Montero, D.S. Multimode Fiber Broadband Access and Self-Referencing Sensor Networks. PhD Thesis. Electronics Technlogy Dpt., Universidad Carlos III de Madrid, Leganés (Madrid), Spain; 2011.

[55] Ingham, J.D., Webster, M., Wake, D., Seeds, A.J., Penty, R.V. and White, I.H. Bidirectional Transmission of 32-QAM Radio Over a Single Multimode Fibre Using 850-nm Vertical-Cavity Half-Duplex Transceivers. Proceedings of the 28th European Conference on Optical Communications (ECOC), Copenhagen, Denmark; 2002.

[56] Gasulla I. and Capmany, J. Simultaneous baseband and radio over fiber signal transmission over a 5 km MMF link. Proceedings of International Topics Meeting on Microwave Photonics Conference (MWP), pp. 209-212, Gold Coast, USA; 2008.

[57] Garcia Larrode, M., Koonen, A.M.J. and Vegas Olmos J.J. Overcoming Modal Bandwidth Limitation in Radio-over-Multimode Fiber Links. IEEE Photonics Technology Letters 2006; 18(22) 2428-2430.

[58] Garcia Larrode, M. and Koonen, A.M.J. Theoretical and Experimental Demonstration of OFM Robustness Against Modal Dispersion Impairments in Radio Over Multimode Fiber Links. Journal of Lightwave Technology 2008; 26(12) 1722-1728.

[59] Garcia Larrode, M., Koonen, A.M.J. and Vegas Olmos, J.J. Fiber-based broadband wireless access employing optical frequency multiplication. IEEE Journal of Selected Topics in Quantum Electronics 2006; 12(4) 875-881.

[60] Garcia Larrode, M., Koonen, A.M.J., Vegas Olmos, J.J. and Ng'Oma, A. Bidirectional radio-over-fiber link employing optical frequency multiplication. IEEE Photonics Technology Letters 2006; 18(1) 241-243.

[61] Raddatz L. and White, I.H. Overcoming the modal bandwidth limitation of multimode fiber by using passband modulation. IEEE Photonics Technology Letters 1999; 11(2) 266-268.

[62] Tyler, E.J., Webster, M., Penty, R.V. and White, I.H. Penalty free subcarrier modulated multimode fiber links for datacomm applications beyond the bandwidth limit. IEEE Photonics Technology Letters 2002; 14(1) 110-112.

[63] Zhao, X. and Choa, F.S. Demonstration of 10-Gb/s transmissions over a 1.5-km-long multimode fiber using equalization techniques. IEEE Photonics Technology Letters 2002; 14(8) 1187-1189.

[64] Pepeljugoski, P., Schaub, J., Tierno, J., Wilson, B., Kash, J., Gowda, S., Wu, H. and Hajimiri, A. Improved Performance of 10 Gb/s Multimode Fiber Optic Links Using Equalization. Proceedings of Optical Fiber Communication and the National Fiber Optic Engineers Conference (OFC/NFOEC), pp. 472-474, Atlanta, USA; 2003.

[65] Raddatz, L., White, I.H., Cunningham, D.G., Nowell, M.C, Tan, M.R.T. and Wang, S.Y. Fiber-optic m-ary modulation scheme using multiple light sources. Proceedings of Optical Fiber Communication and the National Fiber Optic Engineers Conference (OFC/NFOEC), pp. 198-199, Dallas, USA; 1997.

[66] Raddatz, L., White, I.H., Cunningham, D.G. and Nowell, M.C. An experimental and theoretical study of the offset launch technique for the enhancement of the bandwidth of multimode fiber links. Journal of Lightwave Technology 1998; 16(3) 324-331.

[67] Webster, M., Raddatz, L., White, I.H. and Cunningham, D.G. A statistical analysis of conditioned launch for gigabit ethernet links using multimode fiber. Journal of Lightwave Technology 1999; 17(9) 1532-1541.

[68] Sun, Q., Ingham, J.D., Penty, R.V., White, I.H. and Cunningham, D.G. Twin-Spot Launch for Enhancement of Multimode-Fiber Communication Links. Proceedings of Conference on Lasers and Electro-Optics CLEO; Baltimore, USA; 2007.

[69] Stuart, H.R. Dispersive multiplexing in multimode fiber. Science 2000; 289 305-307.

[70] De Boer, M., Tsekrekos, C.P., Martinez, A., Kurniawan, H., Bergmans, J.W.H, Koonen, A.M.J., van den Boom, H.P.A. and Willems, F.M.J. A first demonstrator for a

mode group diversity multiplexing communication system. Proceeding of IEE Seminar on Optical Fibre Communications and Electronic Signal Processing, pp. 0_46-16/5, London, UK; 2005.

[71] Dixon, B.J., Pollard, R.D. and Iezekiel, S. Orthogonal frequency-division multiplexing in wireless communication systems with multimode fiber feeds. IEEE Transactions on Microwave Theory and Techniques 2001; 49(8) 1404-1409.

[72] Shen, X., Kahn, J.M. and Horowitz, M.A. Compensation for multimode fiber dispersion by adaptive optics. Optics Letters 2005; 30(22) 2985-2987.

[73] Patel, K.M., Polley, A., Balemarthy, K. and Ralph, S.E. Spatially resolved detection and equalization of modal dispersion limited multimode fiber links. Journal of Lightwave Technology 2006; 24(7) 2629-2636.

[74] Gu, X.J., Mohammed, W. and Smith, P.W. Demonstration of all-fiber WDM for multimode fiber local area networks. IEEE Photonics Technology Letters 2006; 18(1) 244-246.

[75] Yadlowsky M.J. and Mickelson, A.R. Distributed loss and mode coupling and their effect on time-dependent propagation in multimode fibers. Applied Optics 1993; 32(33) 6664-6677.

[76] Gloge, D. Optical power flow in multimode optical fibers. Bell Labs Technical Journal 1972; 51 1767-1783.

[77] Tatekura, K, Itoh, K. and Matsumoto, T. Techniques and Formulations for Mode Coupling of Multimode Optical Fibers. IEEE Transactions on Microwave Theory and Techniques 1978; 26(7) 487-493.

[78] Djordjevich, A. and Savovic. S. Numerical solution of the power flow equation in step-index plastic optical fibers. Journal of the Optical Society of America: B-Optical Physics 2004; 21 1437-1442.

[79] Zubia, J., Durana, G., Aldabaldetreku, G., Arrue, J., Losada, M.A. and Lopez-Higuera, M. New method to calculate mode conversion coefficients in SI multimode optical fibers. Journal of Lightwave Technology 2003; 21(3) 776-781.

[80] Stepniak, G. and Siuzdak, J. An efficient method for calculation of the MM fiber frequency response in the presence of mode coupling. Optical and Quantum Electronics 2006; 38(15) 1195-1201.

[81] Saleh, B.E.A. and Abdula, R.M. Optical Interference and Pulse-Propagation in Multimode Fibers. Fiber and Integrated Optics 1985; 5 161-201.

[82] Gasulla, I. and Capmany, J. Transfer function of multimode fiber links using an electric field propagation model: Application to Radio over Fibre Systems. Optics Express 2006; 14(20) 9051-9070.

[83] Capmany, J., Ortega, B., Pastor, D. and Sales, S. Discrete-Time Optical Signal Processing of Microwave Signals. Journal of Lightwave Technology 2005; 23(2) 703-723.

[84] Montero, D.S. and Vázquez, C. Temperature impairment characterization in radio-over-multimode fiber systems. Optical Engineering 2012; 51(8) 085001-7.

[85] Hackert, M.J. Characterizing multimode fiber bandwidth for Gigabit Ethernet applications. Proceedings of Symposium on Optical Fiber Measurements 1998; 113-118.

[86] TIA/EIA Fiber Optic Test Procedure (FOTP) 54. Mode Scrambler Requirements for overfilled Launching Conditions to Multi-Mode Fibers; Dec 2001.

[87] Ishigure, T., Sato, M., Kondo, A. and Koike, Y. High-bandwidth graded-index polymer optical fiber with high-temperature stability. Journal of Lightwave Technology 2002; 20(8) 1443-1448.

[88] Montero, D.S. and Vázquez, C. Analysis of the electric field propagation method: theoretical model applied to perfluorinated graded-index polymer optical fiber links. Optics Letters 2011; 36(20) 4116-4118.

Plastic Optical Fiber Technologies

Chapter 6

Step-Index PMMA Fibers and Their Applications

Silvio Abrate, Roberto Gaudino and Guido Perrone

Additional information is available at the end of the chapter

1. Introduction

With the general term "Optical Fibers" it is quite common to refer to a specific type of fibers, in particular Glass Optical Fibers (GOF), that can then be divided into several categories depending on the type of applications they are needed for (communications, sensing, lasing, etc.); but optical fibers are not only glass-based: a wide variety of Polymer-based Optical Fibers (POF), that can be mainly classified based on the specific material and the index profile, exists, for several applications.

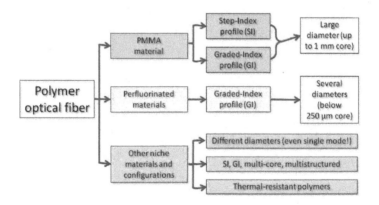

Figure 1. Overview of the different types of POF available.

Two major classes of POF can be identified: Step-Index POF with large core and Graded-Index POF. It is quite common to identify the first type of fiber as POF and the second one as PF-POF (made of perfluorinated material) or GI-POF, however in the following, for sake of clarity, PMMA-SI-POF will be used to address large core step index fibers made of PMMA material. Some other variants exist but are not commonly used, so we will not address them in this chapter.

The use of polymers instead of glass gives certain advantages in terms of mechanical robustness and installation in hostile environments (such as in presence of water or high humidity), so many studies are still in progress to reduce the transmission performance penalty that POF pay with respect to GOF. Since the behavior of the best performing GI-POF are getting very similar to multi-mode GOF, purpose of this chapter is to focus only on PMMA-SI-POF.

This chapter is organized as follows: first, we will give an overview on the fiber itself, describing the material, the production process, the main characteristic; secondly, we will describe components and tools for PMMA-SI-POF handling and using; then, we will analyze their adoption for communication systems and sensing applications.

2. Basics of PMMA-SI-POF

The most widely available PMMA-SI-POF has a core diameter of 980 μm and a global (core plus cladding) diameter of 1mm, while a variant with a diameter of 500μm is gaining interest; however, only the first type of fiber is standardized [1].

The success of 1mm fiber is due to the wide range of applications (Hi-Fi, car infotainment systems, video-surveillance, home networking) and to the interesting mechanical characteristics with respect to GOF. In particular, we can highlight the following main advantages that this type of POF has with respect to other fibers (we will not discuss about all the intrinsic advantages of optical propagation compared with electrical communications, that are maintained moving from GOF to POF):

- High mechanical resilience: the flexibility of the plastic material allows rough handling of the fiber, such as severe bending and stressing, without causing permanent damages. This enables brownfield installation (for example in existing power ducts, being an electrical insulator), also thanks to the 2,2 mm diameter of conventional PMMA-SI-POF simplex cable;

- High mechanical tolerances: the 980 μm core and the 0,5 numerical aperture allow a certain aligning mismatch in connection processes with transmitter and receivers of among fiber spools. This tolerance avoids the use of expensive precision tools for connectorization. Moreover, dust on the fiber ends is less compromising than with small-core fibers;

- Low bending losses: the core diameter also allows a certain bending tolerance. It has been demonstrated [2] that more than 20 bends at 90° with a radius of 14 mm are requested to cause a loss over 5dB for a 1 Gbps transmission system, even if standards foresee 0,5dB for every bend with a bending radius of 25 mm;

- Easy tooling: fiber cut can be made via conventional scissors, and polishing via sand paper, however very simple tools that avoid polishing after cutting exist. Connectorization is fast and easy via crimping or spin connectors, while also connector-less connection via clamping is foreseen in recent transceivers;

- Use of visible sources: the PMMA material works efficiently in the visible wavelength, namely red, green and blue (650 nm, 520 nm and 480 nm respectively). This actually helps unskilled personnel to have a preliminary evaluation of the good functioning of the components (you can actually see the light);

- Ease of installation: the previous characteristics result in a certain ease of installation for unskilled personnel and users, then yielding a consistent reduction in installation time and cost;

- Water resistance: PMMA is also very resistant towards water and salted water. This makes POF suitable for marine applications.

These advantages are reflected in 500 μm PMMA-SI-POF, with the obvious note that alignment tolerances are lower.

In turn, PMMA-SI-POF suffer of high attenuation and low bandwidth; while the attenuation is due to the material, the bandwidth limitations are due to the size of the core and the index profile: in 1 mm PMMA-SI-POF around 1 million modes are propagating in the operational wavelengths.

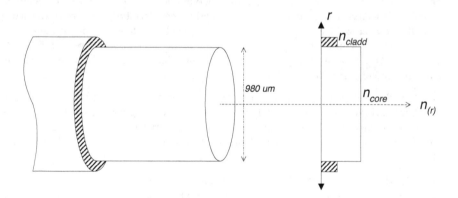

Figure 2. PMMA-SI-POF dimensions and index profile

We can then summarize that PMMA-SI-POF are not to be considered as competitors to GOF, but are rather competitors to copper, with the advantage of being a suitable medium for hostile environments. In Figure 3 it is possible to see a comparison among standard UTP Cat. 5e cable and a PMMA-SI-POF duplex cable.

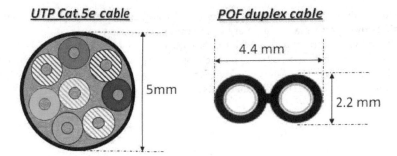

Figure 3. Comparison among a standard UTP Cat.5e copper cable and a PMMA-SI-POF duplex cable. POF cable is smaller and can easily replace copper cable.

2.1. Materials and production processes

2.1.1. Core materials

The most common material for POF is PolyMethylMethAcrylate (PMMA), also known as Plexyglas; it's refractive index is 1,492 and its glass transition temperature is around 105°C. PMMA based POF usually work with visible light (red, green and blue), however the attenuation can be very high (up to 200dB/Km for commercial fibers). Other materials have been investigates: Polystyrene (PS) has an higher refractive index than PMMA (1,59) but its attenuation performances are not expected to be better, so currently no mass production employing this polymer exists; Polycarbonate (PC) has a refractive index of 1,58, is interesting for special applications thanks to its high glass transition temperature (150°C) but its very high attenuation makes it not suitable for telecom/datacom applications.

MMA structure can be seen in Figure 4.

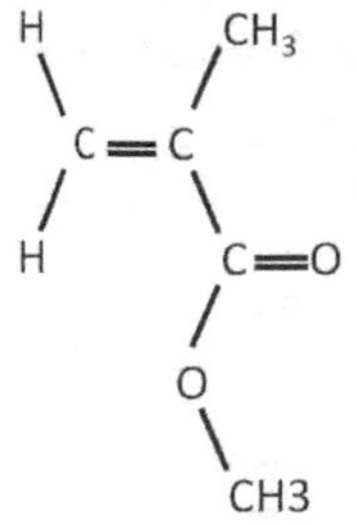

$$
\begin{array}{c}
\overset{\displaystyle H}{\underset{\displaystyle H}{\diagdown}} C = C \overset{\displaystyle CH_3}{\underset{\displaystyle C = O}{\diagup}} \\
\qquad\qquad\qquad O \\
\qquad\qquad\qquad CH3
\end{array}
$$

Figure 4. MMA momomer

2.1.2. Cladding materials

The other main materials for POF are Fluorinated Polymers; they can also be used for the core, since their performances are very interesting in terms of attenuation: in theory it could be comparable with the one achieved for glass fibers, and the refractive index is in the order of 1,42;to date, the best results have been achieved with CYTOP polymer, working at 850 nm and 1300 nm and used for GI-POF. However, from the point of view of PMMA-SI-POF, PF polymers are adopted as cladding materials.

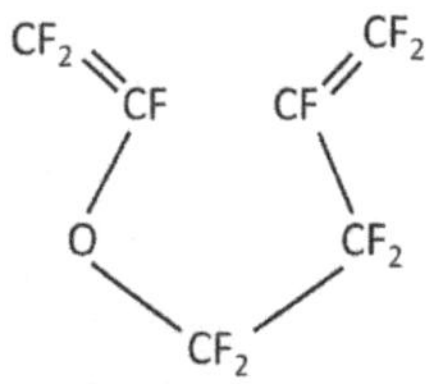

Figure 5. CYTOP momomer

PMMA can be used as cladding material when the core is made with PC.

2.1.3. Manufacturing by fiber drawing

The most well-known method for fiber productions is drawing from a preform, with a proper drawing tower; this method is used for mass production of glass fibers and can be easily adapted to polymer fibers.

A cylinder of polymer (the preform), having the very same structure and refractive indices difference of the fiber we want to draw has to be prepared, usually with an extrusion process; this cylinder has dimensions orders of magnitude bigger compared to the fiber it is meant to generate. The preform is then mounted on top of the drawing tower and heated through a specific furnace to a temperature that makes the polymer starts to soften, so that it becomes possible to reduce its diameter via controlled traction by a take-up winding drum. During the process, the diameter is controlled and it is eventually possible to deposit the coating (however this operation can also be performed in a subsequent phase).

Some variants of the process foresee the preform to be suitable for core drawing only, with the cladding applied subsequently via extrusion.

This way, the length of the fiber that can be obtained is limited by the dimension of the original preform.

With respect to GOF drawing towers, due to the lower melting temperature of polymer with respect to glass, POF towers are lower and also the ovens have a lower working temperature. Also the drawing speed is significantly lower, being in the order of 0,5 m/s while for GOF the conventional production speed overcomes 10 m/s.

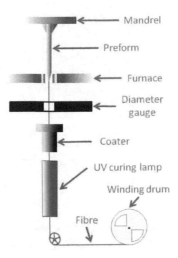

Figure 6. Fiber drawing.

2.1.4. Manufacturing by extrusion

Producing the fiber by extrusion requires the whole process to start from the monomer, that by means of a distillation process is inserted into a proper reactor together with the initiator and the polymerization controller. Once the process is concluded (at a temperature of about 150°C), the polymer is pushed through a nozzle by pressurized nitrogen injections, in order to control the diameter, and the cladding is applied (the cladding is extruded at around 200°C).

The extrusion is quite simple for PMMA-SI-POF, and is the most promising manufacturing process since it is quite cheap and allows continuous production starting from the monomer, thus enabling mass-production.

2.2. PMMASI-POF characteristics

2.2.1. Attenuation

Attenuation is a very important factor in determining the maximum length of a fiber link, and depends on the material properties and the transmission wavelength. The PMMA attenuation spectrum is depicted in Figure 7. It can be seen that, as happens with glass, three transmission windows can be clearly identified, even if with very different attenuation values: around 500 nm, 570 nm and 650 nm, starting from at least 80dB/Km; being in the visible wavelength interval, these windows can be associated to colors, respectively blue-green, yellow and red.

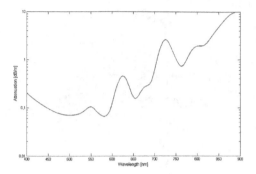

Figure 7. PMMA attenuation spectrum. Different windows can be identified.

The availability of components and the shape of the windows actually suggests to identify the transmission windows as follows: blue (480 nm), green (520 nm) and red (650 nm). Green and blue windows are characterized by the lowest attenuation, in the order of 80dB/Km (together with yellow, in which the attenuation is even lower but there is lack of components, and thus this window will be neglected in the following of this chapter), while in red the attenuation is nearly doubled but where there is a significantly higher availability of components at higher speeds. It has to be mentioned that standards [1] use to define the attenuations as reported in Table 1 for PMMA-SI-POF dubbed of category A4a.2.

Wavelength (nm)	Attenuation (dB/Km)
500	<110
650	<180

Table 1. Attenuation of PMMA-SI-POF according to IEC 60793-2-40 A4a.2

It is then evident that, when dealing with PMMA-POF, transmission length is limited to a few tens or a few hundred meters, depending on the baud-rate.

Given the attenuation of the fiber and the fact that home/office networking is one of the most interesting market for data-communications over PMMA-POF, bending loss becomes a parameter of paramount importance when dimensioning and then installing the system. As previously mentioned, standards foresee 0,5dB for a bend with a radius of 25 mm, but better results have been achieved; Figure 8 shows measured value of extra-losses for 360° bends, when the modal equilibrium is reached.

It can then be said that 0,5dB of extra-loss has to be considered for each 10 mm bend, while there is virtually no extra-loss to be considered when the bending radius exceeds 25 mm.

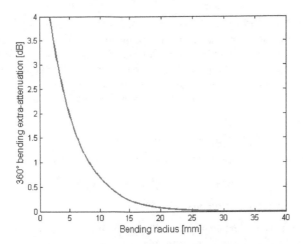

Figure 8. Extra attenuation vs. bending radius

As briefly mentioned, the modal equilibrium condition is important while measuring attenuation: due the multimodality of the fiber, the launching conditions are important especially for short lengths. In order to avoid having length-dependent attenuation measurements (after a certain length the Equilibrium Mode Distribution EMD is naturally obtained), usually two methods are adopted: differential measurement with consistent fiber lengths or the insertion of a mode scrambler at the transmitter side. An example of mode scrambler is reported in Figure 9.

Figure 9. Mode scrambler. Two cylinders with a radius of 21 mm are separated by 3 mm. The fiber is wounded in a 8-shape 10 times around those cylinders. The total attenuation of such an arrangement is about 10dB.

2.2.2. Bandwidth

PMMA-SI-POF are highly multi-modal (in the order of 1 million modes), and in the wavelength regime we consider, for what concerns bandwidth performances, multi-modality is by far the most limiting factor, while chromatic dispersion becomes negligible. It is not target of this chapter to perform a deep theoretical analyses of bandwidth in POF, then we will now focus only on experimental measurements, pointing out the fact that, as GOF, POF have a low-pass characteristic that can be approximated with a Gaussian curve.

A bandwidth measurement technique has not yet been defined in any standard; in literature, we can find results exploiting the following methods:

1. Frequency-domain direct spectral measurement with network analyzers;

2. Time-domain measurement with narrow pulse generation;

3. Optical Time Domain Reflectometry (OTDR).

The most comprehensive results available in literature [3] have been obtained with method 1, while results obtained with the other methods are usually a lot more limited in the length of the link [4], [5].

Frequency-domain measurement setup is quite simple: an electrical network analyzer drives an high-speed laser source connected to the fiber under test, then an high-bandwidth optical receiver closes the loop into the network analyzer, so that a direct bandwidth measurement can be performed. The results shown in Figure 10 are referred to a fiber with a declared NA=0,46. It is evident how POF systems can also be bandwidth limited, since we range from 30 MHz for 100 m of fiber to 9 MHz for 400 m of fiber. Also in this case it is useful to reach the EMD condition to avoid measurement being affected by launching conditions, such as transmitter numerical aperture.

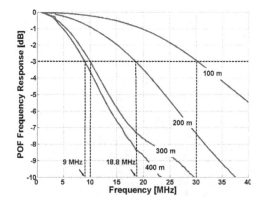

Figure 10. Electrical-to-electrical PMMA-SI-POF response for different link lengths with indication of 3dB bandwidth. Courtesy of the authors of [3].

It is not purpose of this chapter do go into deep analysis of the theoretical aspects of fibers bandwidth, and we suggest to refer to [6] if interested.

2.2.3. Handling, tooling and connectorization

The big advantage of 1 mm POF are due to their easy handling: this does not require expensive equipment and allows do-it-yourself installation; in particular:

- PMMA-SI-POF is robust and flexible, with good bending properties, and thus suitable for careless handling;

- its core dimension and numerical aperture allow certain mechanical tolerances and low sensitivity to contaminations;

- connectorization is easy, requiring simple tools (such as even conventional scissors) and, taken to the extreme, also allows connector-less contact.

Workmanlike connectorization of PMMA-SI-POF foresees the following steps:

- cutting and stripping the fiber with a proper tool, such as in Figure 11;

- inserting the fiber into the chosen connector (different types of connectors can be seen in Figure 12) and locking it (the connectors are usually self-crimping or screw-type);

- putting the connector into a polishing disk (Figure 13) and cleaving by moving the disk on delicate sand paper forming several times a 8-shape.

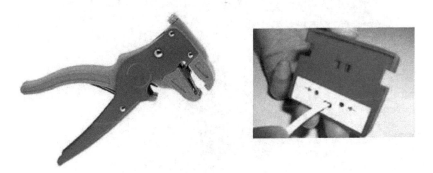

Figure 11. Cutting and stripping tools. On the left, a conventional copper cable stripper; on the right, a proper tool courtesy of Firecomms.

For such a connection, a 1 dB penalty is usually taken into account. Fusion splicing is not available with POF, so splicing is obtained facing to end-connectors into a proper in-line connector, and thus a 2 dB attenuation has to be taken into account.

Figure 12. Different type of 1 mm POF connectors. ST, SMA (2 versions), V-pin. Other type of connectors exist.

It is worth nothing that connectorless installation is gaining real interest since the induced penalties with respect to the previously mentioned procedure can be really negligible if the cutting is made with a certain care. If cutting and stripping is done with tools such as the ones shown in Figure 11, allowing a certain plain cut of the end face, then special transceiver housings such as the Optolock™ (by Firecomms, Figure 14) can be used, simply inserting the fiber into it and then locking.

Figure 13. Polishing disk for 1 mm POF. This disk will be moved forming several times a 8-shape on sand paper for final cleaving.

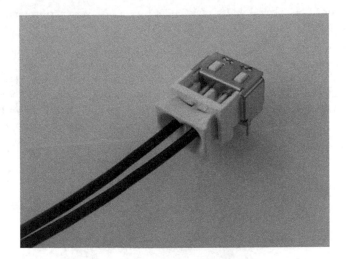

Figure 14. Optoloc™ transceiver housing, courtesy of Firecomms.

2.3. Overview on components

It is not in the scope of this chapter to present a full treatise on optical components, that would deserve a full book itself, so we suggest to consult [7] for this purpose and we will give a very general overview on what type of optical components are available for PMMA-SI-POF applications, given that the most interesting novelties of PMMA-SI-POF components are related to the optical sources only.

2.3.1. Sources

LEDs are the most common optical source to be employed with PMMA-SI-POF. LEDs are available for all the main wavelengths (red, green and blue), and can guarantee high output power and long lifetime. Components with an output power of up to +6 dBm can be found on market, and modulation bandwidths usually are in the order of the tenth of megahertz; thus, they usually are suitable for low-speed transmissions, such as 10 Mb/s, or require complex modulation formats of equalization techniques for higher speeds. Typical linewidth of LED sources is in the order of 40 nm.

A wide variety of red lasers exist, mostly developed of CD and DVD drives and laser pointers; usually, sources developed for such applications hardly meet the speed requirements for data communications but might be suitable for sensing applications. High power edge emitting lasers suitable for high-speeds exist, but not yet available in mass production or for low-cost applications. Vertical Cavity red lasers (VCSELs) are gaining interest since they can achieve interesting performances in terms of bit-rate [14], however low-cost commercial units usually have their peak wavelength at 665 nm, that remains in the red region but experiences a little attenuation penalty with respect to sources working at the optimal wavelength of 650 nm. The spectral width of VCSELs is of course very narrow, and the typical output power is in the range of -5 dBm to -2 dBm.

Resonant Cavity LEDs (RC-LEDs) are gaining increasing interest for communications, since they join the robustness of LEDs with the high bandwidth provided by the resonant cavity. Commercial components work at 650 nm, with a spectral width in the order of 20 nm. Commercial RC-LED have 2 or 4 Quantum Wells (2QW or 4QW); in general 2QW sources are faster while 4QW sources are more powerful. On average, the typical bandwidth of a RC-LED source is in the order of 250 MHz, while the output power goes up to 0 dBm.

For comparison purposes, in Figure 15 and 16 are reported the eye diagrams at the output of commercial low-cost VCSEL and a RC-LED when transmitting 1,1 Gb/s.

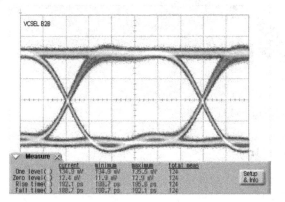

Figure 15. Gb/s transmission, eye-diagram at VCSEL output

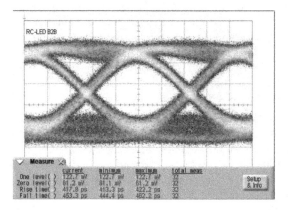

Figure 16. Gb/s transmission, eye-diagram at RC-LED output

As a summary, it is worth reminding that when needing high-speed components, such as VCSELs and RC-LEDs, then working in red wavelength is the only option.

2.3.2. Photodiodes

Typically, silicon photodiodes are used with PMMA-SI-POF. Their highest responsivity is usually around 950 nm, but their efficiency usually remains quite high also at 650 nm; some variants having their best performance at 800 nm exist. The performances decay when working at shorter wavelengths, but the lower attenuation of the fiber in green and blue.

Typical photodiodes have an area of 500 μm, up to 800 μm; considering the fiber diameter of 980 μm, it is quite common to use spherical coupling lenses in the photodiode package for improving coupling efficiency.

Pin structures are the most common to be found on market, but some Avalanche Photo Detectors (APD) can also be found.

2.3.3. Passive components

In the POF world there is not the same variety of passive components as in the GOF world. In particular, it can be said that only POF couplers exist off-the-shelf. The reasons for this lack of components is mainly due to the relatively low market needs. In particular, it can be said that only couplers/splitters exist off-the-shelf, mainly used for measurements setups or sensing applications. Couplers for PMMA-SI-POF are in general quite simple to be produced, mainly starting from the fiber itself: the most common structure foresees to polish two fibers, match and then glue them. It has to be mentioned that such couplers usually exhibit an excess loss in the order of 3 dB (to be added to the 3 dB due to the power splitting).

It is then worth mentioning that, however filtering in the visible regime should be quite common, no filters for PMMA-SI-POF exist. At the same time, no attenuator are available, and the common way to obtain (uncontrolled) attenuation is to insert in-line connectors into a fiber link and then creating an air-gap among the two facing fibers.

3. Data communications with PMMA-SI-POF

Considering attenuation and bandwidth characteristics illustrated in paragraph 2.2 and the performances of the components described in paragraph 2.3, it becomes quite evident that, if we consider the speeds defined by the Ethernet standard, 10 Mb/s systems are mainly attenuation limited, while transmitting at 100 Mb/s and over suffers of severe bandwidth limitations. Communications with PMMA-SI-POF then require the adoptions of mechanisms that are not usual to the optical community but that are widely adopted for example in copper or radio communications, such as multi-level modulation schemes or equalizations. In the following we will rapidly describe the most interesting multilevel modulation formats currently adopted for PMMA-SI-POF transmission, then we will report on the architectures that in literature have demonstrated the best bit rate vs. length results, considering the data-rates defined by the Ethernet standard.

3.1. Amplitude modulations: binary and multilevel

Amplitude modulations are the only formats reasonably applicable to PMMA-SI-POF systems, due to the unavailability of external modulators.

Conventional optical communications adopt On-Off Keying (OOK), that is a binary amplitude modulation, thus transmitting one bit per symbol and that in optics can be simplified switching the source ON when transmitting symbol 1 and OFF when transmitting symbol 0. In recent years more complex modulation formats, able to transmit more bits per symbol, have gained interest also when dealing with single-mode GOF for ultra-high capacity backbone systems. When dealing with PMMA-SI-POF, also due to the absence of proper optical modulators, only direct modulation of the source power can be adopted, thus introducing *Pulse Amplitude Modulation* (PAM).

PAM) consists in transmitting one of M possible amplitude levels (the "symbols") in each time slot. It is a well-known technique outside the fiber optic community, while it has found so far little (if any) application in fiber transmissions. For this reason, we briefly review its basic principle and terminology.

The number of levels M is set to $M=2^{N\_bit}$, where N_bit is the number of transmitted bits per symbol. Being T_s the duration of a symbol, the quantity $D=1/T\_s$ is the number of transmitted symbols per second, also called baud-rate, and the resulting bit rate is $B\_r = N\_bit*D$. The only reason for choosing multilevel is that, for a given available bandwidth B_av (related to the cascade of the transmitter, channel and receiver transfer functions), the maximum data rate that can be transmitted without excessive Inter-Symbol Interference (ISI) increases with the number of levels M. As a rule of thumb, the relation:

$B\_av > 0.7\,D$

should be satisfied to have acceptable ISI level (the constant 0.7 comes from the SDH standard; it can vary a little depending on filter types, without qualitatively affecting the following considerations nevertheless). Thus, for the same available bandwidth B_av, the resulting maximum bit rate increases with N_bit following the relation:

$B\_r\_max < N\_bit * B\_av / 0.7$

When adopting OOK, this means that for example 70 MHz are required for a line-rate of 100 Mb/s, while for multilevel modulations with the same bandwidth 100 Mbaud can be transmitted.

The use of multilevel transmission is very interesting for any bandwidth-limited system. On the other side, the drawbacks are:

- for a given Bit Error Rate and a given receiver noise floor, the required received power (or "receiver sensitivity") increases with N_bit

- the entire transmission channel, from the transmitter to the receiver, should be as linear as possible

- the complexity of the TX-RX pair is clearly increased with respect to binary transmission.

Multilevel transmission is then an appealing approach to improve the maximum bit rate without changing the optical part of the system. This key advantage has to be weighted up together with the previously mentioned drawbacks. In particular:

- regarding receiver sensitivity, for the same total bit rate, the penalty of multilevel compared to binary is equal to 1.76 dB for M=4, 3.93 dB for M=8 and 5.74 dB for M=16, if the receiver bandwidth is properly optimized. Without receiver bandwidth optimization, the penalty is respectively 4.77 dB, 9.03 dB and 12.04 dB. These penalties should clearly be taken into account.

- Regarding POF channel linearity, the only significantly nonlinear optoelectronic device is the LED, while the POF itself and the photodiode are linear to a fairly good approximation. Multilevel POF transmitter should therefore properly compensate for potential LED nonlinearity

- Regarding TX-RX electronic complexity, the cost of high-speed electronics is decreasing so much that there is a rationale to move "logical complexity" from the optical level to the electronic level, by using suitable digital signal processing (using programmable devices such as DSP and FPGA).

PAM has been described in deep since it is one of the options that is being considered for the standardization of 1 Gb/s PMMA-SI-POF systems, however other multilevel formats, such adduobinary [8], [9], [10] can be of interest and easy to be introduced.

Increase of performances could also be obtained using adaptive equalization; this topic is too complex to be fruitfully addressed in this chapter, so we will only mention when in literature equalization has been adopted and we suggest the reader to consult [11] for the theory of equalization.

3.2. Best results available in literature

3.2.1. 10 Mb/s transmission

According to the frequency response depicted in Figure 10 and the rule-of-the-thumb reported in the previous paragraph about the relationship among bandwidth and baud-rate, a conventional OOK modulation at 10 Mb/s could easily overcome, in terms of bandwidth, a distance of 400 m. In terms of attenuation, it makes sense then to use green wavelength due to the lowest attenuation it presence: the lack of fast components is not a limiting factor at this bit-rate. However, overcoming 400 m implies a power budget of over 40 dB, impossible with the best receivers available on market. Thus, we can affirm that at 10 Mb/s the system is attenuation limited.

UTP to POF Ethernet media converters currently available on market usually have a maximum reach in the order 200/250 m. They are mostly obtained by using standard Ethernet chipsets and directly driving the optical source. With the same technique, analog video-surveillance systems are being produced.

The best result available in literature [3] shows the possibility of transmitting 10 Mb/s over a distance of 425 Mb/s, by properly choosing the optical components (for mass production) and introducing Reed Solomon Forward Error Correction (FEC). Ethernet transport over such distances has required to correct the standard at level 1 and level 2, removing the Manchester line-coding (that doubles the line rate with respect to the bit-rate) to adopt a 8B / 10B line coding, and transforming the data stream from bursty to continuous in order to apply the FEC.

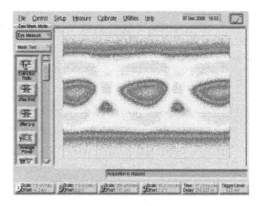

Figure 17. Eye-diagram of 10 Mb/s transmission over 400 m of PMMA-SI-POF, with one intermediate connector. Courtesy of the authors of [3].

3.2.2. 100 Mb/s transmission

Severe bandwidth limitations occur when transmitting at 100 Mb/s: from a power-budget point of view, transmitting in green could target 250 to 300 m, while over these distances the available bandwidth is well below the 20 MHz. This is then the typical case in which multilevel transmission techniques become of paramount importance. Adopting bandwidth-efficient modulation formats can allow, also in this case, the adoption of green components even giver their lack of speed with respect to red components. In fact, the best result available in literature [11] adopts a green LED with a bandwidth of 35 MHz and an average output power of +2 dBm at the transmitter side and a large area photodiode with integrated transimpedence amplifier, with a bandwidth of 26 MHz, at the receiver side, and reaches a distance of 275 m. The authors of the paper have opted for 8 levels PAM (8-PAM), and due to the linearity requirements mentioned in 2.4.1, LED non-linearity compensation has been implemented; even with these techniques, the received eye-diagram after a link in the order of 200 m resulted completely closed, showing that also equalization techniques [12] should be studied in order to recover the signal. In fact, the authors of [11] have adopted adaptive equalization (adaptive to cope with the intrinsic stochastic properties of multimodal dispersion), and the power budget has been increased with the adoption of FEC. In Figure 18 it is shown the eye-diagram of the 8-PAM signal after 200 m of PMMA-SI-POF when LED non-linearity compensation and adaptive equalization are adopted. Moving modulation formats with even more levels would be practically unfeasible for stricter linearity requirements.

It is worth mentioning that, when it is not requested to reach long distances, so that the available fiber bandwidth is bigger, it might be useful to employ red components, faster (such as VCSELs or RC-LEDs) than the ones working in green, and multilevel modulations might be avoided.

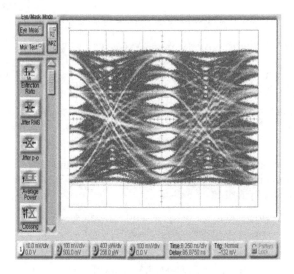

Figure 18. Received 8-PAM signal after 200 m of PMMA-SI-POF, with LED non-linearity compensation and adaptive equalization. Net data rate of 100 Mb/s. Courtesy of the authors of [11].

3.2.3. 1 Gb/s transmission

1 Gb/s transmission over PMMA-SI-POF experiences huge bandwidth limitations, and there is no other chance than using red components and strong equalization. The best results available in literature are due to the POF-PLUS European Project [13], in which it has been shown that in this case complex modulation formats do not give significant advantage with respect to OOK when already equalization is adopted. In [14] it has been shown that with a RC-LED OOK modulated and proper equalization and error correction it is possible to obtain a system overcoming 50 m (75 m with no margin have been obtained). Some little additional margin has been shown in [15] adopting duobinary modulation, a multilevel modulation that has a more complex theoretical background but an easier implementation,

with the current electronic capabilities, than PAM, and is feasible with low cost components. Transmissions over 100 m have been achieved using an edge-emitting laser with an output power of +6 dBm, but such a system cannot be acceptable for practical systems since not eye-safe.

A standardization process is currently going on inside the VDE/DKE initiative, for standardizing 1 Gb/s systems. Since adopting lasers at the transmitter side becomes of interest at this bit rate, then exploiting at most their linearity makes sense, and in fact a solution that adopts Discrete Multi-Tone (DMT) with PAM that adjusts the speed according to the channel performances is currently under investigation [16]: as previously mentioned, PAM vs OOK does not give significant advantages in terms of maximum distance, but in conjunction with DMT inserts in the system rate-adaption capabilities.

3.3. What about WDM over PMMA-SI-POF?

Wavelength Division Multiplexing (WDM) is a very common multiplexing technique adopted for high capacity optical communications with glass fibers; it might appear as an interesting chance with POF as well, but actually it is not a practical solution [17] for high-speed or long-distance applications for the following reasons:

- Array Waveguides (AVG), Mach-Zehender Interferometers (MZI) or Fiber Bragg Gratings (FBG) cannot be used with multimode fibers, so dense wavelength filtering is not possible;

- Red, Green and Blue (RGB) multiplexing is possible but no integrated wavelength splitter exists; experimental units with high insertion losses (5 dB), but in absence of in-line amplifiers this consistently reduces the distance.

- The different performances in terms of attenuation and speed of the components in the three transmission windows would make RGB WDM systems very unbalanced.

In turn, it is possible to say that RGB WDM on PMMA-SI-POF is of interest when low aggregate speeds and short distances are requested; in particular, video systems or medical applications could take advantage of such a technology.

When requiring high speeds and longer distances, the parallel optics approach can be a viable solution, for example for optical interconnects applications [18].

4. Sensing with PMMA-SI-POF

The peculiar characteristics of plastic optical fibers have attracted also the interest in sensing applications, and especially for measuring physical quantities in structural health monitoring [19]. Indeed, using multimode PMMA-SIPOF it is possible to realize fiber based sensing systems that balance costs and performances, since this type of fibers does not require complex machines for splicing and polishing, and makes use of simpler connectors and of visible LED sources. Although several sensing techniques have been described in the literature (and some are described in other chapters of this book), PMMA-SI-POF are best suited for the development of sensors that exploit the variation of the received light intensity with the quantity under measurement, which are the so-called intensiometric sensors, and in this paragraph we will address this technique only.

Typical PMMA-SI-POF intensiometric sensors are based on the variation of: (i) the propagation loss along the fiber (either for local microbending, as for example in [20] and [21], or in distributed form, as in [22]); (ii) the light collected after a free space propagation (as in [23], [24], and [25]); (iii) the interaction through evanescent field tails (as in [26], [27] and [28]). The first two approaches are most often used to measure physical quantities like displacements, vibrations and acceleration, whereas the latter for detecting chemicals.

Intensiometric sensors are conceptually very simple – hence the low cost – because their implementation in principle requires just an LED source and a receiver that acts as a power meter. They are, however, very sensitive to disturbances since any fluctuation in the received power (e.g. due to fluctuations in the source or to fiber degradations) is indistinguishable from actual changes in the quantity under measurement. This sensitivity to parasitic quantities is particularly relevant for long-term monitoring of slowly changing quantities, so in these cases proper compensation techniques using reference sensors [29], or more complex interrogation schemes with signals at different wavelengths [30], must be considered.

Limiting our analysis to the sensors used to measure static or dynamic displacements (vibrations), one of the simplest intensiometric sensors can be realized by facing two fibers along a common axis as in Figure 19. The displacement is measured by exploiting the change of the received power with the separation between the two fiber tips due to the beam divergence form the transmitting fiber (Figure 19 - right). This principle of operation has also been applied in early realizations with glass fibers, but with limitations in the measurement range, unless fiber bundles are used. Despite the simplicity, such a transducer, made using standard step-index 1 mm plastic fibers, has been successfully used to develop a sensing system with working range and accuracy within the typical specifications required for long term crack monitoring in cultural heritage preservation applications [23], [29]. In this case the use of PMMA-SI-POF allowed having most of the advantages of fiber sensors, and above all the impossibility to start fires, without the usual costs and complexities, both in terms of manufacturing and deployment.

Given the propagation loss in plastic optical fibers and the free space attenuation, the distance between the sensor and the interrogators is limited to some tens of meters, but this is typically enough to allow placing the electronics in a remote and safe place. Moreover, if unjacketed fibers are used, the visual impact is dramatically reduced, making the sensing system almost invisible.

An example of the results obtained with sensors arranged as in Figure 19 is shown in Figure 20, where a picture of a sensor mounted across a crack and the readings for a period of 18 months are reported. The data in Figure 20-right are corrected to compensate for the environmental parasitic effects using a "null" (reference) sensor, as reported in [29]. The null sensor is a sensor identical to the others but not fixed to edges of the crack under measure. This is an approach common to most types of the sensors and is effective provided that the reference sensor is exposed to the same kind of disturbances as the measuring sensor; so for meaningful readings, particular care must be devoted to ensure that the two sensors are exposed to the same parasitic phenomena (e.g. temperature, stray light, bending, etc.). The strict correlation between seasonal temperature fluctuations and the crack opening/closing are quite evident from the reported plots.

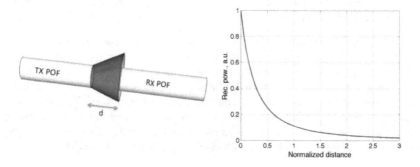

Figure 19. Schematic representation of a POF displacement sensor working in transmission mode (left) and the received power against distance curve (right).

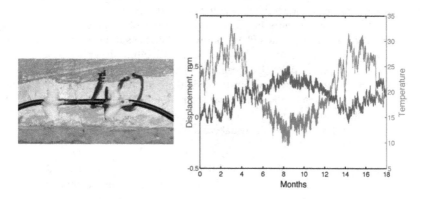

Figure 20. Example of practical POF displacement sensor arranged as in Figure 1 (left), and of the readings of a crack evolution for 18 months, after proper compensation with the null sensor technique as in [11] (right).

A variation of the same working principle is reported in Figure 21, where the light is collected by the receiving fiber after reflection from a target. This configuration can be reduced to the previous one working in transmission mode by considering an image receiving fiber positioned at a double distance and with a lateral offset. The transducer response curve can be modified by changing the sensor geometry (e.g. fiber diameters and separation), but, in any case, it exhibits a maximum that identifies two working regions. The leftmost part of the curve, which is characterized by higher sensitivity, though in a reduced working range, can be used to measure extremely small displacements, such as in high frequency vibrations; however, it requires positioning the sensing head very close to the target. For this reason, in most cases the sensor is arranged to operate exploiting the rightmost part of the curve. This type of sensor can be used both to measure displacements and for non-contact distance measurements.

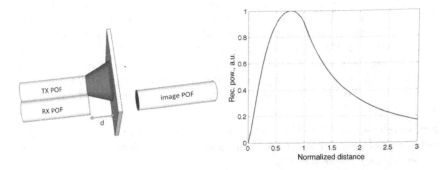

Figure 21. Schematic representation of a POF displacement sensor working in reflection mode (left) and the received power against distance curve (right).

An example of the use to measure displacements is an evolution of the crack monitoring system already shown in Figure 20. Indeed, using the reflection based sensor configuration it has been possible to develop compact transducers having the fiber connections on one side only, as depicted in Figure 4. These new sensors are currently used in a monitoing network deployed inside the chapel hosting the Holy Shroud of Turin in the framework of the Guarini's Project [31], a pilot project devoted to develop new technologies to support the restoration works after the fire that destroyed the Chapel in 1997. In this particular application the POF sensors are integrated within a wireless network to take advantages of both technologies.

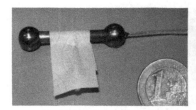 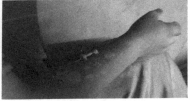

Figure 22. Picture of a crack evolution POF sensor using the principle sketched in Figure 3 (left) and example of application in the Guarini Chapel to monitor a crack on a marble statue in a quite dusty environment (right) [13].

The reflection-based sensor configuration is also particularly well suited for the application of a dual-wavelength compensation technique, which turned out to be much more effective than the null sensor one, though slightly more complex to implement because it requires a dichroic mirror to be inserted in the setup sketched in Figure 22[30]. In this case two signals, at two different wavelengths, are coupled inside the transmitting fiber, then the reference signal is reflected at the fiber tip by a dichroic mirror, while the other wavelength is reflected by the target. This way, the two signals share the same path, hence the same perturbations, except for the sensing region. As for the use in non-contact distance measurements, it is important to highlight that the sensor response depends also on terms that cannot be calculated through theoretical models or may change in time, such as the target reflectivity, so they require continuous characterizations and subsequent calibrations. A sensor for static non-contact distance measurements with response independent from reflectivity changes has been studied in [32], while a calibration technique particularly effective in vibration tests, including cases when the surface has non-uniform reflectivity or non-flat profile, is presented in [33]. An example of a possible application is the mapping of the vibration amplitudes of a printed circuit board under vibration tests. An example of the system setup is pictured in Figure 23.

Recent developments of PMMA-SI-POF displacement sensors include the realization of a possible replacement of conventional crack gage based on sliding plates to measure crack evolutions in two dimensions [34].

Figure 23. Picture of non-contact system for the mapping of the vibration amplitudes of printed circuit boards under vibration tests using the procedure described in [15].

5. Conclusions

In this chapter we have given a general overview of the most interesting applications of optical fibers made of PolyMethylMethAcrylate material, with a core diameter of 980 μm and with Step-Index profile. We have shown that, given the fact that the communication performances are orders of magnitude lower than the ones of the more common single-mode glass fibers, PMMA-SI-POF can address interesting niche markets such as automobile entertainment, local networking, sensing, provided that some complexity is added to the electrical part of the system, while the rules of optical propagation remain unchanged with respect to more common, yet more powerful, optical fibers.

Acknowledgements

The authors of this chapter would like to thanks Stefano Straullu and Valerio Miot for their help in the editing phase, and the partners of the POF-ALL and POF-PLUS European Projects for years of fruitful joint research activities, that have provided for the state-of-the art results in communications over PMMA-SI-POF.

Author details

Silvio Abrate[1], Roberto Gaudino[2] and Guido Perrone[2]

1 Istituto Superiore Mario Boella – Torino, Italy

2 Dipartimento di Elettronica e Telecomunicazioni, Politecnico di Torino - Torino, Italy

References

[1] IEC Recommendation "Optical fibres – Part 2-40: Product specifications – Sectional specification for category A4 multimode fibres", IEC 60793-2-40

[2] A. Nespola, S. Straullu, P. Savio, D. Zeolla, S. Abrate, D. Cardenas, J. C. Ramirez Molina, N. Campione, R. Gaudino, First demonstration of real-time LED-based Gigabit Ethernet transmission of 50 m of A4a.2 SI-POF with significant system margin. 36th European Conference and Exhibition on Optical Communication (ECOC), Turin (Italy), 19-23 September 2010, E-ISBN 978-1-4244-8534-5, DOI 10.1109/ECOC. 2010.5621396

[3] D. Cardenas, A. Nespola, P. Spalla, S. Abrate, R. Gaudino, A media converter prototype for 10Mb/s Ethernet transmission over 425m of Large-Core Step-Index Polymer Optical Fiber. IEEE Journal of Lightwave Technology, vol. 24, n. 12, December 2006

[4] J. Mateo, M. A. Losada, I. Garces, J. Arrue, J. Zubia, D. Kalymnios, High NA POF dependence of bandwidth on fiber length. POF Conference 2003, Seattle (USA), Sept. 2003

[5] E. Capello, G. Perrone, R. Gaudino, POF bandwidth measurement using OTDR. POF Conference 2004, Nurnberg (Germany), Sept. 2004

[6] A. Weinart. Plastic Optical Fibers. Editions Siemens Aktiengesellschaft, ISBN 3-89578-135-5; 1999

[7] O. Ziemann, J. Krauser, P. Zamzow, W. Daum. POF Handbook 2nd Edition. Springer; 2008

[8] A. Lender, Correlative digital communication techniques. IEEE Tran. Commun. Vol. COM-12 pp 128-135, 1964

[9] J. Proakis, M. Salhei. Digital Communication. McGraw Hill; 2007

[10] S. Abrate, S. Straullu, A. Nespola, P. Savio, D. Zeolla, R. Gaudino, J. Ramirez Molina, Duobinary modulation formats for gigabit Ethernet SI-POF transmission systems, International POF Conference, Bilbao (SPA), Sept. 2011

[11] D. Cardenas, A. Nespola, S. Camatel, S. Abrate, R. Gaudino, 100 Mb/s Ethernet transmission over 275 m of large core Step Index Polymer Optical Fiber: results from the POF-ALL European Project. IEEE Journal of Lightwave Technology, vol. 27, n. 14, July 2009

[12] H. Meyer, M. Moeneclaey, S. Fechtel, Digital communication receivers: synchronization, channel estimation and signal processing. Wiley-Interscience; 1997

[13] POF-PLUS European Project, www.ict-pof-plus.eu

[14] A. Nespola, S. Straullu, P. Savio, D. Zeolla, J. C. Ramirez Molina, S. Abrate, R. Gaudino, A new physical layer capable of record gigabit transmission over 1 mm Step Index Polymer Optical Fiber. IEEE Journal of Lightwave Technology, vol. 28, n. 20, 15 October 2010

[15] S. Straullu, A. Nespola, P. Savio, S. Abrate, R. Gaudino, Different modulation formats for Gigabit over POF. ANIC 2012, Colorado Springs (USA), June 2012

[16] KD-POF, http://www.kdpof.com/Papers_files/kdpof_demo_1Gbps.pdf

[17] O. Ziemann, L. Bartkiv, POF-WDM, the truth. POF Conference 2011, Bilbao (SPA), Sept. 2011

[18] S. Abrate, R. Gaudino, C. Zerna, B. Offenbeck, J. Vinogradov, J. Lambkin, A. Nocivelli, 10Gbps POF ribbon transmission for optical interconnects. IEEE Photonic Conference IPC 2011, Arlington (USA) Oct. 2011

[19] K. Peters, Polymer optical fiber sensors—a review, Smart Mater. Struct., vol. 20, 17 pages, 2011

[20] A. Vallan, M.L. Casalicchio, A. Penna, G. Perrone, An intensity based fiber accelerometer, IEEE International Instrumentation and Measurement Technology Conference (I2MTC), pp. 1078-1082, 2012

[21] A. Kulkarni, J. Na, Y. J. Kim, S. Baik, T. Kim, An evaluation of the optical power beam as a force sensor, Opt. Fiber Technol., vol. 15, pp. 131-135, 2009

[22] S. Liehr, P. Lenke, M. Wendt, K. Krebber, M. Seeger, E. Thiele, H. Metschies, B. Gebreselassie, J.C. Munich, Polymer optical fiber sensors for distributed strain measure-

ment and application in structural health monitoring, IEEE Sensors Journal, vol. 9, pp. 1330-1338, 2009

[23] S. Abrate, G. Perrone, R. Gaudino, D. Perla, European Patent n. EP 1630527

[24] M. Olivero, G. Perrone, A. Vallan, S. Abrate, Plastic optical fiber displacement sensor for cracks monitoring, Key Engineering Materials, vol. 347, pp. 487-492, 2007

[25] G. Perrone, A. Vallan, A low-cost optical sensor for noncontact vibration measurements, IEEE Trans. Instr. Meas., vol. 58, pp. 1650-1656, 2009

[26] J. Vaughan, C. Woodyatt, P.J. Scully, Polymer optical coatings for moisture monitoring, European Conference on Lasers and Electro-Optics, 2007

[27] E. Angelini, S. Grassini, A. Neri, M. Parvis, G. Perrone, Plastic optic fiber sensor for cumulative measurements, IEEE International Instrumentation and Measurement Technology Conference (I2MTC), pp. 1666-1670, 2009

[28] S. Corbellini, M. Parvis, S. Grassini, L. Benussi, S. Bianco, S. Colafranceschi, D. Piccolo, Modified POF Sensor for Gaseous Hydrogen Fluoride Monitoring in the Presence of Ionizing Radiations, IEEE Trans. Instr. Meas., vol. 61, pp. 1201-1208, 2012

[29] M.L. Casalicchio, A. Penna, G. Perrone, A. Vallan, Optical fiber sensors for long- and short-term crack monitoring, IEEE Workshop on Environmental, Energy, and Structural Monitoring Systems, EESMS 2009, pp. 87-92, 2009

[30] A. Vallan, M.L. Casalicchio, M. Olivero, G. Perrone, Assessment of a dual-wavelength compensation technique for displacement sensors using plastic optical fibers, IEEE Trans. Instr. Meas., vol. 61, pp. 1377-1383, 2012

[31] M.L. Casalicchio, D. Lopreiato, A. Penna, G. Perrone, A. Vallan, D. Lopreiato, POF sensor network for monitoring the Guarini Chapel, Proc. of the POF Conference, 2011

[32] M.L. Casalicchio, A. Neri, G. Perrone, D. Tosi, A. Vallan, Non-contact low-cost fiber distance sensor with compensation of target reflectivity, IEEE International Instrumentation and Measurement Technology Conference (I2MTC), pp. 1671-1675, 2009

[33] A. Vallan, M.L. Casalicchio, G. Perrone, Displacement and acceleration measurements in vibration tests using a fiber optic sensor, IEEE Trans. Instr. Meas., vol. 59, pp. 1389-1396, 2010

[34] M.L. Casalicchio, M. Olivero, A. Penna, G. Perrone, A. Vallan, Low-cost 2D fiber-based displacement sensor, IEEE International Instrumentation and Measurement Technology Conference (I2MTC), pp. 2078-2082, 2012

Chapter 7

Efficiency Optimization of WDM-POF Network in Shipboard Systems

Hadi Guna, Mohammad Syuhaimi Ab-Rahman,
Malik Sulaiman, Latifah Supian, Norhana Arsad and
Kasmiran Jumari

Additional information is available at the end of the chapter

1. Introduction

Polymer optical fibers (POFs) are in a great demand for the transmission and processing of optical-based data communications compatible with the Internet, which is one of the fastest growing industries in automobile and domestic industry. Other industry such as aviation and maritime have also taking the advantages of POF. POFs become an alternative transmission media replacing copper cable for future shipboard networks. A proposed POF based technology over submarine network for multimedia data transmission, measurement system, navigation, sensors and several applications. As shown in Figure 1, the system is able to transmit a number of signals represent a different data transmission (such as video, audio, etc) using a WDM based network (refer to Figure 1).

In this chapter, we proposed a wavelength division multiplexing (WDM) system over POF due to the rapid increase of traffic demands [2]. WDM is the solution that allows the transmission of data in onboard the ship over more than just a single wavelength (color) and thus greatly increases the POF's bandwidth.

2. Fiber optic onboard ship

The utilization of optical fiber as major data communication media onboard ship especially on naval combatant ships is not a new discovery [1,2]. Equipments such as communications system, radars, navigation system, combat management system, platform monitoring systems and LAN network have used fiber optic to transfer high rate data within equipment or as main data communication backbone. For instant, the platform control and monitoring system onboard ship is using dual redundancy Fiber Data Distribution Integration (FDDI) system to command, control and monitor the platforms onboard the ship. An example of FDDI architecture is shown in Figure 2. This FDDI architecture topology is glass fiber based that capable to transport high density data over long distance because the backbone is covering the entire ship. Numbers of commercial ships are also using FDDI topology as it is a proven system available commercially in the market [1-4].

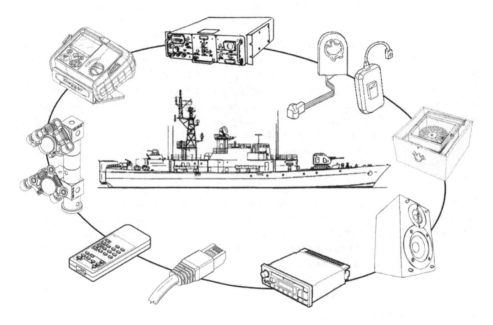

Figure 1. WDM-POF based network over novel system has been propose to ensure the high quality data transmission and communication system

In this chapter, the novel optical splitter/coupler based polymer optical fiber (POF) was successfully designed for the infotainment and data communication system over POF on-ship. The optical splitter consists of a single input port and N of output port (N=2,3,4,....). In principle, the bidirectional splitter performs two operational functions; either signal coupling (in multiple P2P direction) or signal splitting (in P2M direction). Thus, the usage of WDM onboard ship will become a new frontier in optical network.

This fiber optic onboard ship system is the most updated and promising innovation that will revolutionize in-vehicle data communication system which all data can be simply sent in visible light format rather than in electrical format at high speed data transmission. Data communication system is such all-in-one communication media system and the latest trend onboard ship network in which many appliances such as navigation system, platform surveillance & monitoring system, damage control & fire fighting system, onboard infotainment & training, sensors and many other appliances can be integrated *via* a WDM-POF. With this WDM-POF-based technology, all data such can be processed with environment-friendly LED conversion and low-cost multiplexing and filtering method besides the fact that it can extend the number of appliances in car interior. This invention enables simple POF cabling system for delivering each optical data as POF is the most updated cabling technology replacing conventional copper wire for short-haul communication. The advantages offered by POF over copper wire; economical installing cost, enabling eco-friendly LED conversion, Electromagnetic Interference (EMI) immunity, no grounding necessary, avoiding sparks, resistance to heat and vibration, lighter material, and narrow bending radius.

During the implementation of this project, several research activities to improve the efficiency of the system has been conduct. Temperature plays a significant role which can influence the performance of the data communication system in POF-based onboard ship. The characterization test was carried out is to determine the performance of the device in the test bench network. In the meantime, the fabricated splitter has been compared to other commercial one, in term of their performance; splitting ratio and power loss. An experiment has been set up in SPECTECH lab, Universiti Kebangsaan Malaysia, to evaluate the survivability of the device in environmental condition with varied temperature. Besides, the aim of experiment is to observe the temperature stability of the device while performing splitting/coupling function. The variation of temperature from 30 °C to 125 °C was exposed directly to the device.

In response to feedback from industries, the thermal aging experiment was undertaken to evaluate the durability of the device in very high temperature environment. The experiment was carried out within 9 hours while the device was exposed to high temperature at 105 °C. An analysis was made to observe the device performance with the variation of temperature. Several graph was plotted to analyze power loss and coupling ratio in varied temperature.

3. Results and discussion

In this study, single line POF is used to carry multiple wavelengths using WDM technology taking the advantage of its cheaper materials and fragility. Four different wavelengths are used to connect LAN connections, telephone line, surveillance cameras and central video/audio entertainments network throughout the ship for access by the user. The controller and server for ship LAN and surveillance cameras is at Machinery Control Room (MCR) that located at deck 1 aft of the ship. This is also the location of damage control and fire fighting headquarters onboard. The telephone PABX and central video/audio entertainment network controller is at Main Communication Center located at the centre of the ship on deck 1. The systems are also able to be monitored and controlled from the bridge located at 01 deck where the ship is navigated or from the combat Information Centre (CIC) where the ship warfare tactical information and status is collected, displayed, evaluated, disseminated and controlled for decision by the Commanding Officer.

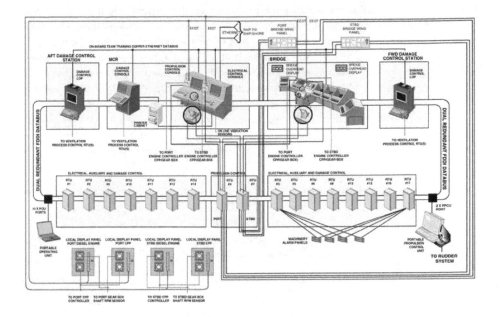

Figure 2. L3 Dual Redundant FDDI for Ship Control and Monitoring architectural network

The CCTV will provide surveillance and monitoring from flood, fire or unauthorized entrance of the high value compartments onboard. The LAN will enable ship staff to access all administration and orders, manuals, publications, maintenance requirements and training document from offices, common area and cabins. The central video/audio entertainments network is providing the ships' crews with central entertainment such as ship's live radio, movies and news broadcasted throughout the ship. The controller is placed at ship's main broadcast & recreation centre. The suggested backbone topology throughout the ship is as shown in Figure 3.

Each deck are interconnected to form a Dual Redundant POF-WDM (DRePOF-WDM) backbone arranged as one ring that interconnected to the equipments and end user devices. The backbone is arranged in mesh topology via an Optical Add Drop Multiplexer (OADM) which acts as optical switches. These switches will be able to be controlled and monitored at MCR, CIC or bridge for redundant connection through the backbone to ensure survivability and interconnectivity of the network. The connection [5] is shown in Figure 4. The devices need for this system is: fiber couplers, Multiplexer, Demultiplexer, Optical Add Drop Multiplexer (OADM) and POF's switches.

Figure 4 as shown above indicates overall arrangement of the system from the backbone to the equipments and the end users located on the various decks onboard the ship. On each deck, equipments and users in the rooms or compartments is linked to the DRePOF-WDM backbone topology using WDM sequenced by time division multiplexing TDM *via* a transceiver. The multiple different signals enter and exit from the devices onto the single wavelength data streams are done by passive devices multiplexer and demultiplexer. Many transmitters with different lights colour are used to carry single information. For example, red light with 650nm wavelength modulated with LAN signal while blue, green, and yellow lights carry image information, radio frequency (RF), and video signal, respectively. As shown is Figure 4, WDM is the first passive device required in WDM-POF system and it functions to combines optical signals from multiple different single-wavelength end devices onto a single fiber [6-7]. Conceptually, the same device can also perform the reverse process with the same WDM techniques, in which the data stream with multiple wavelengths decomposed into multiple single wavelength data streams, called demultiplexing.

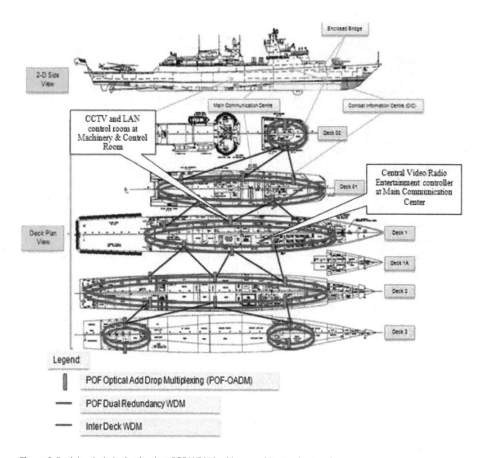

Figure 3. Deck-by-deck dual redundant POF-WDM backbone architectural network

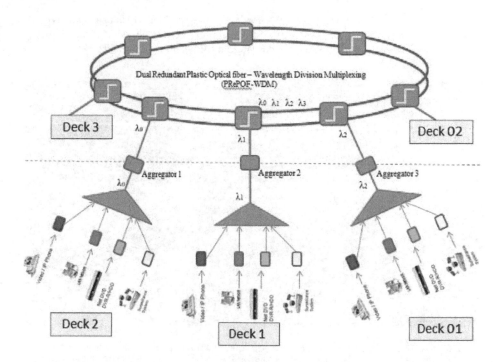

Figure 4. Connection from DRePOF-WDM backbone to each deck and equipments

During the development of the onboard project, several research activities to improve the efficiency of the system has been conduct. The characterization test was carried out is to determine the performance of the device in the test bench network.

3.1. Design and characterization of POF splitter

3.1.1. First generation

The first generation of low-cost fused taper (LFT™) splitters is initially demonstrated as novel innovation in optical splitter technology particularly for POF since it is fabricated via handmade fusion technique that is performed by handwork skill associated with simple tools; candle and metal rather than biconical fused taper. The fabrication method is cost-effective and less time-consuming (11 minutes per unit).

In comparison, the first generation of LFT™ splitter is more cost-effective than other POF-based commercial splitter e.g. Diemount™ grinded splitter, Harz-optic™ splitter, Industrial Fiber Optic™ (IFO) Fused Splitter and many others. The high costs of these commercial splitters are mainly due to the fabrication method that is complicated and implemented with fabrication machine that expensive. For LFT™ splitters, new handwork fusion method lead to low fabrication cost of splitter. The price for IFO™ fused splitter which uses same type as LFT™ splitter cost around USD110 while LFT™ splitter is only cost at ~ USD20. Figure 5 shows the price comparison between the commercial splitter and LFT™ splitters.

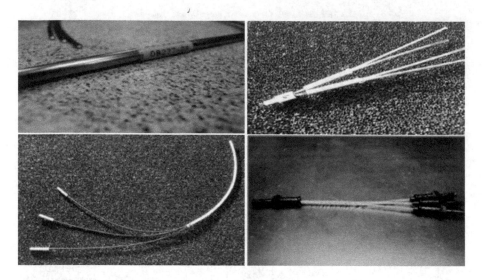

Figure 5. The price comparison between the commercial splitters (a) Industrial Fiber Optic™ (IFO) Fused Splitter (b) Diemount™ grinded splitter, (c) Harz-optic™ splitter, and (d) LFT™ splitters which cost at USD110, USD90, USD50 and USD20, respectively

3.1.2. Second generation

The second generation of LFT™ splitter is the successor of poor-performance fused splitter (first generation). The splitter is remain fabricated through handwork fusion technique. However, the procedures of fabrication method is changed with minor modification whereby the method include a new step particularly for the purpose of fusing the polymer fibers. As shown in Figure 6(a), second generation of LFT™ splitter is designed to have small area of POF imperfection, in which the length of fused and tapered fibers is reduced below 4 cm. The multimode step-indexed *polymethylmethacrylate* (PMMA) POF having a core diameter of 1 mm is used for splitter fabrication. Besides, polyvinyl chloride (PVC) is another material that used as jacket for insulating input and output fiber ports of fused splitter.

In the splitter, the tapered structure is the most critical region in producing low-loss and excellent power-splitting device. The structure has to be designed and fabricated having high fusion degree, in which all POFs are completely fused and coupled so that the wavelength of interest can pass through the coupling region with low power deviation and excellent power-splitting ratio. Therefore, no twisting effetcs are present in tapered region since the twisted spiral fiber is refined via fusion process. Figure 6(b) shows the cross-section of highly fused region in 3 × 3 or/and 1 × 3 tree coupler (splitter). Through fused and pulled region having a cross-section as depicted in Figure 6(b), the optical power input is coupled to each fiber output port with excellent power-splitting ratio. In the other word, one third of power capacity is distributed to every single of output fiber port.

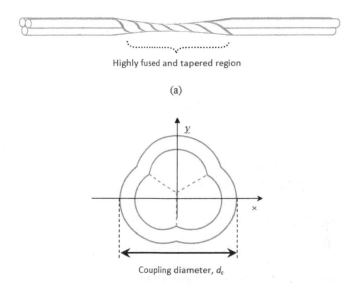

Highly fused and tapered region

(a)

Coupling diameter, d_c

Figure 6. New schematic design of (a) highly-fused taper structure in the center of fiber bundle and (b) cross-section of fused region in fused 3 × 3 biconical coupler

Basically, the term of '*fusion*' defines the act or procedure of liquefying or melting by the application of heat. The maximum temperature required to ensure POFs reach melting point is 85°C [6]. In general, the technique includes four processes; fiber bundle configuration, fabrication of spiral fiber, fusion and fiber tapering. Among these process, fusion is new step that firstly demonstrated in fabrication method for the second generation splitters.

a.   Fiber bundle configuration

b.   Fabrication of spiral fiber

c.   Fusion

d.   Fiber tapering

Since the length of tapered fiber is reduced below 6 cm to minimize area of POF imperfection. An experimental characterization was undertaken on the relationship between the length of tapered and optical loss to observe a possible range of tapered length enabling low-loss power splitting. Figure 7 shows the relationship between the length of tapered and optical loss. Figure 8 shows the relationship between coupling ratio and the length of tapered. These results are essential in determine excellent dimension for the fabrication of second generation of LFT™ splitter. Coupling ratio is a parameter that indicates fusion characteristic in fused fibers. The ideal coupling ratio is 0.33 for each output port of splitter. The coupling ratio of 0.33 for each port shows that each fiber has been fused completely to be as a new single core.

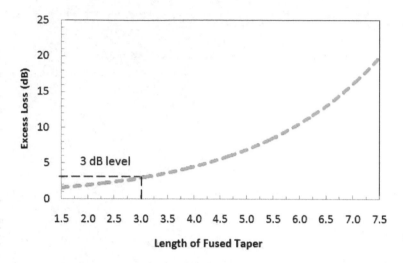

Figure 7. The excess loss of 3 × 3 coupler with range of tapered length vary from 1.5 cm to 7.5 cm; low excess loss < 3 dB occur in coupler in the range of tapered fiber length of 1.5 – 3.0 cm.

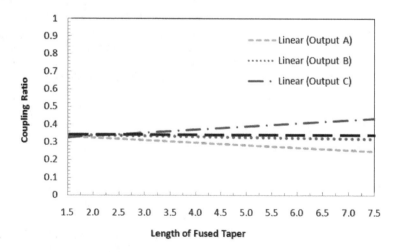

Figure 8. The coupling ratio of 3 × 3 coupler with range of tapered length vary from 1.5 cm to 7.5 cm; the coupler has good coupling ratio (~ 0.33) for each port within the range of 3 cm to 1.5 cm

From the graph, it is indicated that low optical loss < 3dB presents in tapered fiber length range of 1.5 – 3.0 cm. Furthermore, the fused and tapered fiber has good fusion characteristic in the range of 1.5 – 2.0 cm since the coupling ratio of each output fiber reach ~ 0.33 within this range. It is found that 1.5 cm is the minimum length required for fused input fiber to be suited into a small channel having ~1 mm diameter in DNP connector. Therefore, the range of 1.5 – 2.0 is selected for excellent dimension of fused and tapered length in order to permit low-loss power splitting and homogenous splitting ratio. Figure 9 (a) shows the structure of fused and tapered output fiber featured in the second generation of LFT™ splitter, in which the diameter of POF cross-section decrease to ~1 mm that fabricated through modified handwork fusion technique.

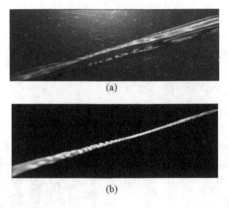

(a)

(b)

Figure 9. The features of (a) novel highly fused tapered having short taper length and plane surface (without twisting effect) and (b) conventional fused taper having long taper length and ripple surface (with twisting effects)

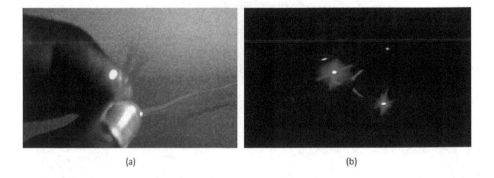

(a)   (b)

Figure 10. The results of experimental optical injection with 650 nm light source; (a) for the first generation of splitter and (b) for the second generation of LFT™ splitter.

As shown in Figure 10 (a), when the only one fused input port is injected with red LED transmitter having 650 nm, it is observed that each output port emits high-intensity red light. In comparison, as shown in Figure 10 (b), in the past experimental injection test, each output port of the first generation splitter emits red light with low power intensity except one output fiber among them. The power splitting with high intensity shows that the second generation of fused splitter is able to perform low-loss optical data splitting.

For the first generation splitters, as shown in Figure 11, the insertion loss of each output port is high which the range is 10 - 20 dB. In contrast to the first generation splitter, the second generation splitters perform with low insertion loss since each output fiber has insertion loss varying from 4 dB to 17 dB.

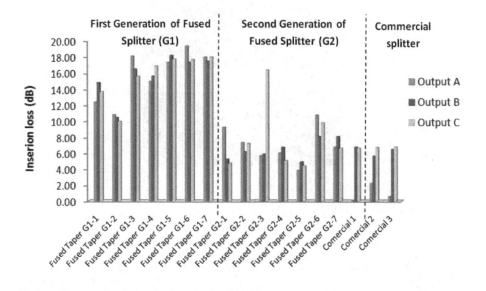

Figure 11. The comparison for insertion loss of each output fiber

Figure 12 shows the result for excess loss of the first and the second generations of LFT™ splitter and commercial splitter. The result shows that the excess loss of the second generation splitter is lower than the first generation; this means that the performance of low-cost fused splitter has been improved effectively.

3.2. Temperature effect experiment

In the experiment, temperature of hot plate was increased by 5 °C to reach stable condition. Figure 13. shows the influence of temperature variation T from 30 °C to 125 °C on output power P_o for the splitter.

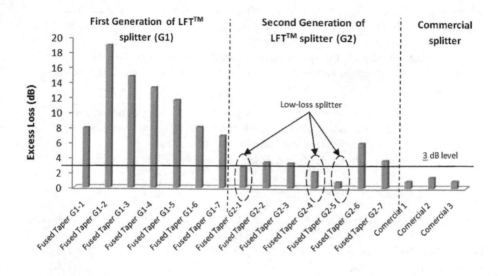

Figure 12. The comparison for excess loss

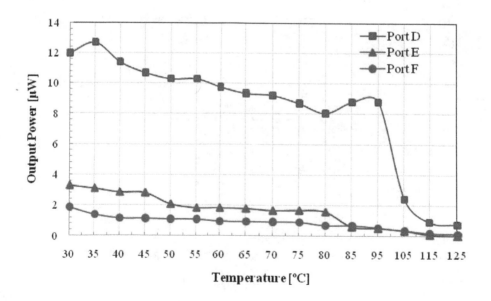

Figure 13. The relationship between temperature variation (30 °C to 125 °C) and output power for Low cost POF splitter

As shown in Figure 14, in each fiber port, output power decreases with respect to temperature rise. The type of fused polymer splitters were completely damaged when heating temperature increased $T = 125$ °C. The temperature point at 95 °C can thus be defined as damage threshold because the splitter loss temperature stability at this point. Figure 14. shows Excess loss variations as function of temperature for the splitter in bidirectional power injection.

As shown in Figure 15, the excess loss increase gradually with temperature increase. In this case, the splitter has temperature stability while maintaining their performance until $T = 100$ °C. Figure 15. shows temperature dependence of coupling ratio for the splitter in their throughput and cross-coupled fiber ports in bidirectional light guide propagation.

3.3. Thermal aging experiment

Figure 16. shows the durability of the Low cost POF splitter within 9 hours at fixed temperature $T = 105$ °. The graph shows that the splitter has high temperature stability within 9 hours when the splitter was exposed to very high temperature. The result shows that the splitter has high durability.

Figure 17. shows the durability of the Low cost POF splitter in term of output power in μW within 9 hours at fixed temperature $T = 105$ °.

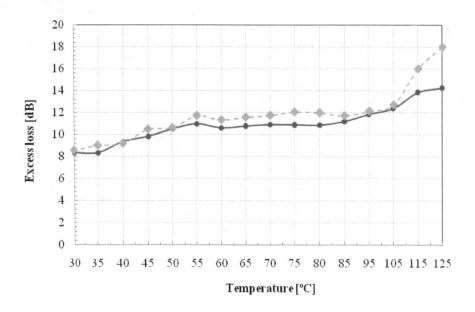

Figure 14. Excess loss variations as function of temperature for the splitter in bidirectional power injection

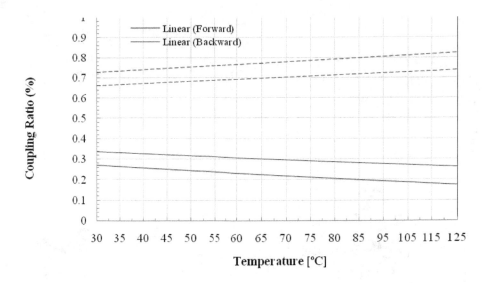

Figure 15. Excess loss variations with temperature increase for the splitter

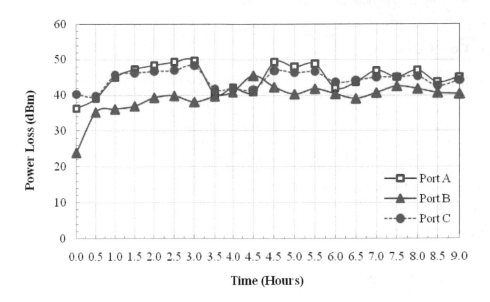

Figure 16. The relationship between heating time and power loss of the splitter

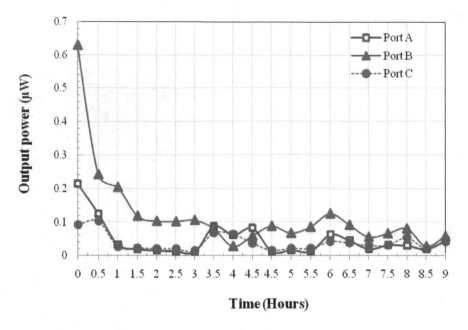

Figure 17. The relationship between heating time and power loss of the splitter

4. Conclusion

In conclusion, the Wavelength Division Multiplexing application over the Polimer Optical Fiber was used for data transmission onboard ship system. The network has been designed via dual redundancy POF-WDM interconnected deck-by-deck using mesh topology, introducing the design philosophy of Dual Redundant POF-WDM (DRePOF-WDM) backbone network. OADM acts as switches is used to make redundancy circuits [5, 8]. Four different wavelengths has been used to connect the overall equipments throughout the ship. This system is very promising hence the payback of less overall ship's weight and therefore will improve the speed and less fuel consumptions of the ship for future new build or ship embarking life extension program. The efficiency related to the temperature effect and thermal aging has been observed in order to optimized the onboard ship communication network. Any system or equipment to be fitted onboard can use this existing DRePOF-WDM backbone.

Acknowledgements

This research has been conducted in Computer& Network Security Laboratory, Universiti Kebangsaan Malaysia (UKM). This project is supported by Ministry of Science, technology and Environment, Government of Malaysia, PRGS/1/11/TK/UKM/03/1. All of the handmade fabrication method of POF splitter, LFT^{TM} splitter and also the low cost WDM-POF network solution were protected by patent numbered PI2010700001.

Author details

Hadi Guna, Mohammad Syuhaimi Ab-Rahman, Malik Sulaiman, Latifah Supian, Norhana Arsad and Kasmiran Jumari

Universiti Kebangsaan Malaysia, Selangor Darul Ehsan, Malaysia

References

[1] Gopalkrishna, D.H., S.R. Muthangi, Vinayak, A. Paulraj. FDDI-A high speed data highway for warship system integration 1991, MILCOM '91.

[2] Gotthardt, M.R., K. Kathiresan, G.H. Campbell, J. Fluevog. Shipboard Fiber Optic Transmission Media (Cables). AT&T Bell Labs, AT&T Federal Systems and AT&T Networks Systems. CH2681-5/89/0000-0235 IEEE (1989): 0235.

[3] Baldwin, C., J. Kiddy, T. Salter, P. Chen, J. Niemczuk. Fiber Optic Structural Health Monitoring System: Raugh Sea Trials of the RV Triton 2002. Ocean MTS/IEEE.

[4] Jessop, Clifford N. All-Optical Wavelength Conversion in a Local Area Network, Report to United States Naval Academy Trident Scholar Committee, May 2006.

[5] Mohamad Syuhaimi Ab-Rahman, Mohamad Najib Mohamad Saupe, Kasmiran Jumari. Optical Switch Controller Using Embedded Internet Based System 2009. International Conference on Computer Technology and Development.

[6] Imran Ahmed, Hong Wong and Vikram Kapila. Internet-Based Remote Control using a Microcontroller and an Embedded Ethernet, Proceeding of the American Control Conference: 2004.

[7] Hong Wong and Vikram Kapila. Internet-Based Remote Control of a DC Motor using an Embedded Microcontroller 2004, Proceeding of the American Society for Engineering Education annual Conference and Exposition.

[8] Alam, M.F., M. Atiquzzaman, B. Duncan, H. Nguyen, R. Kunath. Fibre-optic network architectures for on-board radar and avionics signal distribution. IEEE Radar Conference, Washington, DC, 7-12 May, 2000.

Permissions

The contributors of this book come from diverse backgrounds, making this book a truly international effort. This book will bring forth new frontiers with its revolutionizing research information and detailed analysis of the nascent developments around the world.

We would like to thank Sulaiman Wadi Harun and Hamzah Arof, for lending their expertise to make the book truly unique. They have played a crucial role in the development of this book. Without their invaluable contribution this book wouldn't have been possible. They have made vital efforts to compile up to date information on the varied aspects of this subject to make this book a valuable addition to the collection of many professionals and students.

This book was conceptualized with the vision of imparting up-to-date information and advanced data in this field. To ensure the same, a matchless editorial board was set up. Every individual on the board went through rigorous rounds of assessment to prove their worth. After which they invested a large part of their time researching and compiling the most relevant data for our readers. Conferences and sessions were held from time to time between the editorial board and the contributing authors to present the data in the most comprehensible form. The editorial team has worked tirelessly to provide valuable and valid information to help people across the globe.

Every chapter published in this book has been scrutinized by our experts. Their significance has been extensively debated. The topics covered herein carry significant findings which will fuel the growth of the discipline. They may even be implemented as practical applications or may be referred to as a beginning point for another development. Chapters in this book were first published by InTech; hereby published with permission under the Creative Commons Attribution License or equivalent.

The editorial board has been involved in producing this book since its inception. They have spent rigorous hours researching and exploring the diverse topics which have resulted in the successful publishing of this book. They have passed on their knowledge of decades through this book. To expedite this challenging task, the publisher supported the team at every step. A small team of assistant editors was also appointed to further simplify the editing procedure and attain best results for the readers.

Our editorial team has been hand-picked from every corner of the world. Their multi-ethnicity adds dynamic inputs to the discussions which result in innovative outcomes. These outcomes are then further discussed with the researchers and contributors who give their valuable feedback and opinion regarding the same. The feedback is then collaborated with the researches and they are edited in a comprehensive manner to aid the understanding of the subject.

Apart from the editorial board, the designing team has also invested a significant amount of their time in understanding the subject and creating the most relevant covers. They scrutinized every image to scout for the most suitable representation of the subject and create an appropriate cover for the book.

The publishing team has been involved in this book since its early stages. They were actively engaged in every process, be it collecting the data, connecting with the contributors or procuring relevant information. The team has been an ardent support to the editorial, designing and production team. Their endless efforts to recruit the best for this project, has resulted in the accomplishment of this book. They are a veteran in the field of academics and their pool of knowledge is as vast as their experience in printing. Their expertise and guidance has proved useful at every step. Their uncompromising quality standards have made this book an exceptional effort. Their encouragement from time to time has been an inspiration for everyone.

The publisher and the editorial board hope that this book will prove to be a valuable piece of knowledge for researchers, students, practitioners and scholars across the globe.

List of Contributors

Danish Rafique
Photonic Systems Group, Tyndall National Institute and Department of EE/Physics, University College Cork, Dyke Parade, Ireland
Now with Nokia Siemens Networks, S.A., Lisbon, Portugal

Andrew D. Ellis
Photonic Systems Group, Tyndall National Institute and Department of EE/Physics, University College Cork, Dyke Parade, Ireland

Claudio Risso, Franco Robledo and Pablo Sartor
Engineering Faculty – University of the Republic, Montevideo, Uruguay

Mitsuru Kihara
NTT Access Network Service Systems Laboratories, Nippon Telegraph and Telephone Corporation, Japan

Aurenice M. Oliveira
Michigan Technological University, U.S.A

Ivan T. Lima Jr.
North Dakota State University, U.S.A

David R. Sánchez Montero and Carmen Vázquez García
Displays and Photonics Applications Group (GDAF), Electronics Technology Dpt., Carlos III University of Madrid, Leganés (Madrid), Spain

Silvio Abrate
Istituto Superiore Mario Boella – Torino, Italy

Roberto Gaudino and Guido Perrone
Dipartimento di Elettronica e Telecomunicazioni, Politecnico di Torino - Torino, Italy

Hadi Guna, Mohammad Syuhaimi Ab-Rahman, Malik Sulaiman, Latifah Supian, Norhana Arsad and Kasmiran Jumari
Universiti Kebangsaan Malaysia, Selangor Darul Ehsan, Malaysia